GEOGRAPHY AND RURAL DEVELOPMENT

Edited by

R.N.P. SINHA
Department of Geography
Faculty of Science
M.S. University, Baroda

CONCEPT PUBLISHING COMPANY, NEW DELHI 110059

ISBN-13: 978-81-7022-423-5

Edition 2016

Published and Printed by

Concept
Since 1974

Concept Publishing Company Pvt. Ltd.
A/15-16, Commercial Block, Mohan Garden
New Delhi-110059 (India)
T: 25351460, 25351794, *F*: 091-11-25357109
E: publishing@conceptpub.com, W: www.conceptpub.com
Editorial Office: H-13, Bali Nagar, New Delhi-110 015, India.

CONTENTS

LIST OF TABLES

LIST OF FIGURES

LIST OF CONTRIBUTORS

Dr. Anil Kumar Tewari
Reader & Head, Deptt. of Geography, University of Rajasthan, Jodhpur (Rajasthan).

Dr. (Smt.) Anjana Desai
Head, Deptt. of Geography, Gujarat University, Ahmedabad (Gujarat).

Dr. Bhikhubhai C. Parekh
Vice-Chancellor, The M.S. University of Baroda, Baroda (Gujarat).

Dr. Chintamani Gupta
Reader in Geography, The M.S. University of Baroda, Baroda (Gujarat).

Dr. C.S. Rathod
Prof. & Head, Deptt. of Cooperation, The M.S. University of Baroda, Baroda (Gujarat).

D.A. Bhatt
Geography Teacher, Nava Rachna School, Baroda (Gujarat).

Dr. Dinesh C. Pandey
Lecturer, Department of Geography, Kumaun University, Nainital (U.P.).

Shri F.V. Yavagal
Lecturer, Deptt. of Geography, Karnataka University, Dharwad (Karnatak).

Dr. Hari Mohan Saxena
Head, Deptt. of P.G. Studies in Geography, Government College, Sriganganagar (Rajasthan).

Shri Jawwad Ali
Research Scholar, Deptt. of Geography, P.P.N. Post-Graduate College, Kanpur University, Kanpur (U.P.).

Dr. K.K. Sharma
Lecturer in Geography, Govt. College, Ajmer (Rajasthan).

M.G. Bhasin
Head, Deptt. of Geography, D.B.F. Dayanand College of Arts & Science, Solapur (Maharashtra).

Shri M.N. Vaishnav
Chief Conservator of Forest, Baroda (Gujarat).

Dr. Ram Bali Singh
Reader in Geography, University of Gorakhpur, Gorakhpur (U.P.).

Dr. R.C. Srivastava
Lecturer, Deptt. of Geography, P.P.N. Post-Graduate College, Kanpur (U.P.).

Dr. Ram Lochan Singh
Vice-Chancellor, Meerut University, Meerut (U.P.).

Dr. Ram Nandan P. Sinha
Prof. & Head, Deptt. of Geography, The M.S. University of Baroda, Baroda (Gujarat).

Dr. Ramlal Parikh
Vice-Chancellor, Gujarat Vidyapeeth, Ahmedabad (Gujarat).

Shri S.B. Deshmukh
Lecturer in Geography, Shivaji University, Kolhapur (Maharashtra).

S.L. Hiran
Lecturer in Geography, M.L.V. Government College, Bhilwara (Rajasthan).

Shri S.M.H. Rizvi
Lecturer in Geography, The M.S. University of Baroda, Baroda (Gujarat).

Dr. Shrikant Pandey
Lecturer in Geography, Degree College, Ghosi, Distt. Azamgarh (U.P.).

Shri Suchit Sur
Lecturer, Deptt. of Geography & Applied Geography, North Bengal University, Darjeeling (West Bengal).

Shri Sanat Mehta
Finance and Planning Minister, Govt. of Gujarat, Ahmedabad (Gujarat).

INTRODUCTION

THE MAIN purpose of the symposium was to draw the attention of planners and administrators towards the significant role of geographical skill and expertise in the sphere of rural development planning. Therefore, in my welcome address, I underlined the role of geography, an integrative and synthesising science, in the comprehension and solution of the meta-problems (a system of problems) of integrated rural development. The non-recognition by the authorities concerned of the vital role of geography calls for long overdue change. The geographic fraternity should also project their competence in this field of endeavour.

Inaugurating the symposium, Sri Sanat Mehta, the Finance Minister of Gujarat, recognised the important role of geography in the various fields of development and called upon the geographers to widen the scope of geographical study and remove the misunderstandings about the subject from the people's mind and create a situation for its legitimate recognition. Prof. Ramlal Parikh emphasised geographer's stress on total profile of man in the spatial context which enables us to view human problems in its totality, and this is what is required in the context of rural development. Use of geographical method of horizontal distribution can save the development process from causing social disparity. Prof. R.L. Singh in his keynote address viewed rural development on the broader canvas of spatial planning and rural habitat transformation. He pointed out how geography's new core concept of 'socio-spatio-structural environmentalism' has developed from the call from the society for the solution of its future problems. Prof. B.C. Parekh, in his Presidential address traced the evolution of the idea of the Western model of development and exhorted the geographers to think about the Western model, about what is our own tradition that we wish to retain, and how is it possible to absorb the elements of westernisation which are absolutely crucial, how to retain our heritage which we cherish and to reject that which we are ashamed of. He wanted the symposium to trigger this debate which is long overdue.

Part II comprises five papers presented on the theme — Land-use and Rural Development. In the first paper "Land-use Components and Land-use Efficiency in a Peripheral Region" based on village level analysis of Pharenda Tehsil of Gorakhpur District, adjoining the Nepal border and peripheral to Gorakhpur, Kanpur and Saryupar Plain, Dr. Singh and Pandey attempt to identify some salient characteristics of spatial distribution of land-use components to indicate certain trends in temporal variation of these components during a period of three decades, and to evaluate land-use efficiency using the ranking coefficient method. They have found that there exists a gradual south to north gradient in the distribution of land-use components as well as land-use efficiency in the study area. In the second paper, Sri Pandey focuses his attention on crop combination regions as a basis for crop planning in the broader frame of rural development. He is of the view that crop combination regions reflect the optima for the type and number of crops which can be produced in a particular micro-spatial region, and has observed that mono-cultural and bi-crop cultural regions are more likely to be benefited by the technological revolution in agriculture. But, how the technological innovations may sometimes adversely affect cropping pattern has been illustrated by Sri Rizvi and Bhatt with a case study of Amod Taluka in the District of Bharuch (Gujarat). They have shown how the introduction in the area of high yielding varieties of cotton led to the incidence of pests and diseases resulting in a sharp decline in the cultivation of cotton giving place to pulse crop *tur (arhar)*, thus sounding a caution in introducing new technologies in agriculture without proper study of the local environmental peculiarities. Dr. Gupta, thus, advocates the study of land-use on the village level as a prelude to planning the development of rural areas. But how efficiently this land use survey can be carried out with the help of remote sensing technique is the subject-matter of the paper by Sri Deshmukh.

Part III is given to the geographical study of the rural economic activities with a view to bringing out rural development. The activities touched upon are social forestry, rural industrialisation, quarrying and cooperative banking. Sri M.N. Vaishnav highlights the role of social forestry in the rural development of Gujarat. Social forestry, according to him, is the practice of forestry through the people, by the people, for the people and has the immense potential to benefit especially the weaker sections of the society through self-help, self-employment, and through meeting the minimum needs of fuel, fodder, small timber and fruits. It is a mechanism that can marry land and labour without the least strain on the environment and with the least inputs of finance and technical knowhow. Gujarat, he hopes, is sure to provide leadership in this regard. Sri Bhasin pleads for rural industrialisation for the generation of rural change and, therefore, rural development, and underlines the task of geographers to identify probable rural industries in particular areas and to help in integrating these in the broader regional plan frame. This is followed by Sri Hiran's study of the impact of sand-slab mining on the rural landscape and life style of the people on the Bijolian plateau of Rajasthan. He also highlights some of the problems which need solution for proper rural development. Prof. Rathod outlines the role of cooperation in the rural development of Gujarat, and how the knowledge of geography can be fruitfully utilised in the process. He wishes geographers especially to educate the masses, through the medium of maps, the value of cooperation and to bring out the topographical diversity and nature of accessibility to help the location of rural banks.

Part IV comprises two papers dealing with the role of transport and marketing in the process of rural development. Dr. Saxena, in his paper on rural markets from the perspective of development has studied the pattern of rural markets and marketing system in the State of Rajasthan and has come to the conclusion that the development of rural markets will definitely induce economic growth of a region and also help bring about equitable spatial equilibrium of economic development. Dr. Srivastava investigated the matter more intensively and, taking the Patha area of Bundelkhand as a case, analyses the transport network, explains the spatial distribution of market facilities and villagers' visit pattern, highlights the problem of transport and marketing facilities and finally suggests a sound network of roads and market centres to obtain an optimal spatial organisation of these facilities for the development of this backward area.

In Part V the papers relate geography straight to the various aspects of integrated rural development. Dr. Tewari deals with the spatial aspects of rural development schemes in the Indian desert and comes to the conclusion that the desert region of India can be developed not only by irrigation agriculture but also by organisation and upgradation of non-crop economies and establishing new settlements with varied functions. Sri Yavagal studies the Bijapur district of Karnataka from the point of view of location and allocation of central places with a view to inducing integrated area development of the district. He identifies 48 central places, groups them into a four-fold hierarchical order, identifies the functional gaps and allots the required functions to the proper centres to minimise the gaps. Sri Sur, after a brief appraisal of the concept of spatial planning, takes up the case of Siliguri Sub-division of Darjeeling District (West Bengal) for the application of the idea. He identifies the focal points and ranks them in four orders of hierarchy and then demarcates the hinterlands of the focal points. Ultimately he comes to the conclusion that the utility of this method is "not total but partial" and geographers' concept of 'distance' and 'complimentarity' must count for more. However, he thinks, development of such individual spatial units and focal points will ultimately bring about the total development of a given area. Dr. (Mrs.) Desai also first discusses the concept of development and then tries to view the problem of development of the Dangs District of Gujarat, from that angle, and suggests some measures for the growth suited to that backward tribal district. Dr. Sharma closes the discussion with focusing attention on the negative aspects concomitant with rural development, that of rising criminality as a result of the dismantling of the traditional institutions and non-inception of the modern institutions, thereby creating a gap. He tests this hypothesis with the help of field studies in Bhagwanpura village of Rajasthan. He pleads with the planners and others to make provision for the prevention and control of criminogenesis.

R.N.P. SINHA

PART I
INAUGURAL PROCEEDINGS

1

ROLE OF GEOGRAPHERS IN RURAL DEVELOPMENT

R.N.P. SINHA*

COMING TOGETHER of the luminaries from different streams of thought and action on a common platform of rural development is not without significance. We live today in a world with complex problems not just a single problem, but what we may call, metaproblems, a system of problems, each related to others. Specially, when we talk of integrated rural development, we envisage to face the metaproblems of an economic, social, cultural and political nature. And once the development process is set into motion, we are expected to face the consequences of change affecting all these facets of community life. They all interact with each other in a reflective way. One of the vexing problems facing the leaders of the developing world is that of achieving economic progress without drastically disturbing the distinctive cultural, religious and social characteristics of their people. If the geographers are now to prove successful in grappling with the problems of rural development, then surely we need to join forces with our fellow social scientists, economists, economic and social historians, cultural anthropologists and the social and behavioural psychologists, as well as with the natural scientists and technologists like the geologists, botanists, padolists, agricultural and irrigation engineers etc. Geography, being an integrative discipline itself, can serve as the best platform for coming together of people of various shades and colours working towards the common goal of development.

* Welcome address by the Director of the Symposium.

In the field of integrated rural development, which is distinctly inter and multidisciplinary in character there is vital work to be done and geography is uniquely well suited to provide its central organising and synthesising focus.

But, unfortunately, the worth of Geographers have not been recognised by the persons involved in the decision-making processes. The results are not far to seek. The Five Year Plans have failed to achieve in ample measure the policy objectives. This is because, among other things, the plans conceived in national and State capitals are often out of touch with conditions on the ground and particularly with local environment and resource contrasts. Detailed diagnostic survey and mapping of local conditions are essential before prescription of any remedial measure for the ills of the Community. And this is the task of geographers who have been neglected, especially, at the regional and local levels.

This is because the people outside the geographic profession do not understand the nature and scope of modern geography. They still carry the image of geography as a compendium of place names, rivers and mountains. They are ignorant of the fundamental change that this discipline has undergone in recent years. Geographers are no longer content simply to describe man's world. They ask not only where things and people are on the globe, but also why they are there and how they got there, and most recently, where they out to be, which is surely impinging on the exercise for planning.

But the fault does not lie on the other side only. Geographers themselves have been shy to advertise their concern and competence in this area. That is why we have not been successful in getting other people to share our views of our competence. Perhaps it is time we were more prepared to put ourselves forward to submit ourselves to the test of experience.

We shall feel satisfied if our deliberations in the Symposium draw the attention of the administration and the public towards the role of geographers and involve them in the processes of rural development, and if the geographers come forward and accept the challenge.

2

GEOGRAPHY AND RURAL DEVELOPMENT
An Administrator's Viewpoint

SANAT MEHTA*

WHEN I read the subject of this symposium, I tried to stir up the minds of a few of my friends. They were all a little unsure about the relationship between geography and rural development. This is because, as rightly said by Prof. Sinha in the welcome address, the word 'geography' has still remained in the peripheral use of the old syllabus, even though actually it has gone beyond the concept of the old times. When I was trying to understand the subject, I just picked up the dictionary to find out what the dictionary had to state about geography. One dictionary said that geography is the science of earth: the land, sea, climate, people and products. We all know that we have transformed ourselves, our society and economy, which are based on land, labour, capital, energy and raw materials. And if you go into greater detail about these various factors, which is the basis of our development, you can understand that geography is at the root of all these aspects.

Now, coming to the main subject, you all know that large government resources over the next five years are going to be spent in rural areas. After so many experiments of various types, we have now finally decided to implement the Integrated Rural Development Programme. If you look at the history of all the programmes of rural

* Inaugural address by the Finance and Planning Minister, Govt. of Gujarat.

development, we started with Community Development and we thought that once some *Panchayat Ghars* are constructed, some roads are constructed in rural areas, automatically the rural areas will become developed. In this the pioneer of Community Development, S.K. De, was finally disappointed and he thought that perhaps the Green Revolution might change the plight of the rural poor. But we have seen before our eyes that even in areas where the Green Revolution has taken place, in Surat District, where perennial water is available, the condition of the rural poor has changed very little. So we are again disappointed. Then we thought, let us try a new approach. And the new approach was to evolve a scheme for specific Target Groups. So we took up this type of schemes for small and marginal farmers. Again we observed that benefits did not percolate down to the intended segments. After this, we thought that perhaps the Area Approach might be more fruitful. Thus we evolved schemes like D.P.A.P. and other such schemes. But in spite of all these schemes the rural scene has not changed visibly. And so finally we came to the programme which is called the Integrated Rural Development Programme.

The term 'Integrated Rural Development' means that we have to look from the point of view of total development; development of the people, development of the area, development of the resources available, and from this achieve the integrated rural development. The more I think about our integrated rural development, I feel that geographers can play a very important role in this programme.

Take the case of Gujarat. If somebody says that this scheme has evolved in Delhi and Gandhinagar and it has to be implemented here, it is bound to fail, because each block, each taluka is different, each district is different, each state is different. Schemes meant for one region must therefore take factors specific to that region into account, without which it would be a failure. We work out our resources inventory, and try to understand different blocks, with their different requirements, different data, different potential, and different resources. Take the example of Kachchh. Recently we studied the development of Kutch area to work out the type of integrated rural development programme most suited to this region. Apart from being an arid zone, we found people extensively engaged in various handicrafts. Even without any government help their turnover of handicrafts in the region was about Rs. 29 crore. It is these details that would enable us to formulate the best possible strategy for the development of the region.

Again you go to another area. You will find different types of areas each with different types of conditions. Then you will immediately understand the issues which we are going to face in each region.

Now, take another example. Gujarat at present is suffering from a slow growth of agriculture. It is lower than the all-India average. But then we are also that part of the country where the industrial growth is much faster. So what happens in the totality? That the agricultural growth is very slow on the one hand, but on the other, in the secondary sector and the tertiary sector the growth is very fast; the slow growth in one sector is compensated in other sectors. Now, you want to increase the growth of agriculture in this part of the country, and unless you do it the face of the people is not going to change, because most of our industrialisation is taking place only in a ribbon starting from Ahmedabad and going to Bombay between the National Highway and the Railway line. If you look from the total development point of view, you will see that Gujarat is second in the country as far as industrial development is concerned. But if you look to the development of the people, 40 per cent of rural people live below poverty line. It means that unless you understand each area, the geography of each region, the problem cannot be solved. The State Government have decided that if you want to remove poverty, if you want to help the development of the tribal belt which is a major area of poverty, you have to open up the belt. If you have to open up the belt you have to understand the geography. And to understand the geography we have taken the decision that whatever may be the attitude of the Government of India about the opening of the new highway, Government of Gujarat will spend Rs. 35 crores to build another parallel highway from Shamlaji to Vapi. Now, this decision can be considered as a geographical decision. It will not be only a road linking Shamlaji and Vapi, it will actually go to open up the area. The decision is not taken from the point of view of physical nature but the decision has been taken from the point of view of development and dispersion of industry. Now, if you want to have dispersion of industries you have again to understand the location, terrain, the people, and their habits. Unless you understand all these, your development plan will not be perfect and will not achieve the results.

Another thing that I want to mention is social forestry. Now everybody says that Gujarat has considerably developed social forestry, and you will be surprised to know about this from the World Bank Report of 1981. If you read the report you will get the impression that Gujarat is a separate country. On the page of social forestry they have placed a box item in which two States are mentioned: Gujarat and Korea. Now, Korea is a country and Gujarat is a part of India. And you see, how well social forestry has done in Gujarat. Look at the problem of social forestry. It is really a new feature which was not conceived by any scientist. You will be surprised to know that in the last one year the farmers took by themselves 5000 of eucalyptus saplings for developing social forestry. This year the demand went to ten crores and for 1982-83, we have made plans to supply fifteen crores of saplings in Gujarat. This is not for forest development alone. The farmer themselves are going to have this kind of forestry, farm forestry, which we are now trying to evolve.

Now look to another aspect. The other day I was in Bhavnagar, and I was discussing how this farm forestry can be integrated within Integrated Rural Development Programme. The statisticians and economists say that if you develop a close spacing high density variety of eucalyptus with a little water available in the field, you can have a net profit of Rs. 7,000 per acre. So, now the Government is thinking of taking the help of banks. We are arguing with the Government of India for taking up social forestry as a part of integrated rural development so that it can be a banking programme. This can change the whole pattern of life. We can also say that it can change the whole geography because in a district like Bhavnagar nearly twenty thousand acres of land have gone under eucalyptus. You can see what change, ecological and climatic, will take place in a new area. So we are thinking and discussing how we can provide a programme of farm forestry. It may be lucina, it may be eucalyptus, so that a marginal farmer who has one acre or one and a half acres of land, while growing three rows of eucalyptus and one row of lucina on the borders of the fields, can get net increase every third year of Rs. 1800 to 2000. And if a bad year comes, Irrigation Commission says, that one out of three is bad, the marginal farmer becomes indebted because he cannot survive in that distress. Now this is the area open to us. We are trying to persuade my Forest Department that the department must have an up-to-date inventory of the wasteland and the fallow land. In each survey number for the given type of soil, which kind of aforestation or farm forestry can be taken up? They should be totally computerised so that any time we ask a question we can get a reply. For each Taluka we should know that this much fallow land is there. And then we can understand these different aspects.

Now take the example of wind. We are now developing wind source of energy. Wind energy is a very important development. Now if you want to understand it, know about it, you have to know what is the wind velocity in Gujarat. Fortunately, statistics says that only two States, West Bengal and Gujarat, are in the zone of high wind velocity. This means that it can open up a new programme, if you have basic data about the wind velocity. Now immediately another aspect comes to mind. We are discussing how to improve the plight of the 20,000 families engaged in salt manufacturing. Because we are a coastal State, we have a large coastline. We can have wind velocity utilised in the manufacture of salt. Well, you cannot provide electricity in that area because out of twelve months, for four months, it is flooded with sea water, five to six feet deep. And when the water recedes, they will go, take the engine, develop the salt pans, and start production. As soon as the monsoons come they shift to the villages. All this development is going to be linked up with the Integrated Rural Development Programme.

But all these years we understood Geography as only something which indicates the towns etc. All political systems are also based on geography. You know, I would represent a particular Taluka. No body indeed asks me what type of Taluka it is? What type of natural constraints are there? What kind of difficulties are there, what are the habits of the people there? Not these. He would only say that certain person represents a certain constituency. So our political system is based on geography. But to us geography means only the districts, the talukas, the divisions, nothing more than that. And you know, geography is the only department of science in which even the students, the young people, are not interested, even though the parents are enlighted; they sometimes bring home *National Geographic*, which is a very outstanding magazine, and people like to keep it on the table so that it looks so prestigious; nothing more than that. When you see that magazine, and when you see our geography textbooks for the initial secondary and higher secondary stages, you find the difference between the real geography and what is taught in our schools at present. I am very happy that you have decided to have a symposium not only on geography but on geography and integrated rural development. I wish that more and more fruitful discussions are held about the real concepts, and basic data are accumulated and provided. Then perhaps we may have the right concepts about the relationship between geography and integrated rural development. This is a very interesting subject.

The other day we were studying with various persons as to what we can do with the Bhal area of Gujarat. You know, geography gave us the understanding that, this is an area where agriculture is not possible at all. We found out from statistics that 22,000 hectares of land are growing *Babul* trees, and now we are working out that *Babul* is a natural growing species in that area, and we can go along that line and create some power generation in that area out of that *Babul* wood. These are the relations that we have to take into account. If you do not understand the area properly then, I would say, no proper development can take place in area like Kachchh because the land-use there has serious problems. The original or primitive knowledge is such that we were deciding simply by some preliminary indication that such an area is a good area and such an area is a bad area. But if we go into details, then we can find out many different types of developments. If today or tomorrow wind energy is developed in this area, in the whole coastal area life will be different because wind can generate electricity, and wind velocity is available. If there was no wind velocity at all the wind energy will not develop.

Similarly, the other day there was a conference of the miners in Gujarat, where I submitted that underground coal gasification is possible in North Gujarat and electricity can be generated. The mining experts told me, "Sir, this is a distant dream. You know, we do not like the distant dreams. This is not possible in the immediate future. This will be possible only after ten years." I replied to the mineral experts, "You are right. As a political leader I would not take interest in such things because I know my life is for five years." Why should I worry whether after five years underground coal gasification will develop or not. But if you look from the point of development this is not so. Within six months time this idea has developed so fast that a team of scientists went to the U.S.S.R., Belgium and France. They came with rich data. Not only that, the other day, the then Prime Minister went to France and they signed a protocol regarding the working of underground coal gasification. And now Government of India has decided that finance is not a question at all. Any amount of money can be made available if we develop this new technology of underground gasification.

All these years our knowledge regarding the geography of coal deposits was that at 2,000 metres the coal was not useful at all. We could not get it out at all. But now the technology has changed. Now people say that even at a depth of 2,000 metres we can burn the coal inside and take out the gas, and after the gas comes out we can separate it, the gas can go for power generation and liquid may go for Amonia. Kalol is the richest area for this lignite. Water is abundant there because it is rich with underground water. Technology has also developed. About 100 wells have been drilled there, and so every inch of land below is known as to what kind of strata is there? what is existing there and so on. So the people from Belgium and France working together for a pilot project say that Kalol is the most suitable area for underground coal gasification.

Now, this is the type of knowledge, this is the type of data, this is the type of thing that we find for the geographers. I hope that this disposition will also stir the minds of geographers, and they will think that they have a very big role to play. I think a yeoman service has been done by the organisers by organising this symposium and I wish that after this symposium, some areas, particularly in this part of the country, are marked and identified which will be studied in detail. And I assure you as Planning Minister that all kinds of help will be given to these kinds of studies in this part of the country because it is the unfolding of science. The whole world is changing. Fortunately or unfortunately, I always feel that India will have to work for some many year on some split level. We may have to work with an old form of economy and old form of technology and at the same time with the most up to date and the most modern technology. These two things will go together for a long time. So your task will be to give us the knowledge about the past as well as to indicate what can happen in the future. If such data are available for each area in Gujarat, I think planning can become much more meaningful, resources can be utilised much more rationally and efficiently. We may be able to know our world more properly.

Somebody told me that Mrs. Gandhi has opened up a new Department of Oceanography and when, I was in Delhi, I called on the Secretary of Department of Oceanography, Mr. Saigal, a very enlightened man. He immediately said that our State can play a leading role as far as the development of oceans is concerned. He said that the University can become a toss player. It was one of the reasons that I had a discussion with the V.C. of Bhavnagar. I told Mr. Dhruva that you can develop some areas with the help of Salt Research Institute and the University and you can have some kind of training and education in oceanography. This is going to be a very important area for us in the coming ten years because offshore drilling requires a lot of knowledge of oceanography. At present we have the knowledge only of drilling, exploration and production. All the rest of services required in the field of oceanography are being obtained from abroad at heavy cost in foreign exchange.

These are the different areas which are coming under geography. Geography is a vast area. When I was trying to hear the Director of the symposium, Prof. Sinha, I thought that he will put up an optimistic note. Now I hope that henceforth in this Symposium geography will not feel any inferiority that it is not being recognised. Nothing can be recognised unless you want them to be recognised, and I wish that you will create that situation.

3

STRATEGY OF SPATIAL PLANNING AND RURAL HABITAT TRANSFORMATION

R.L. SINGH*

Emerging Conceptual Frame

EVERY DISCIPLINE has its own way of analysing man in relation to his environmental setting, particularly in the context of resource appraisal, development and the future prospects. Moreover, every branch of knowledge considers the abstraction, perception, and the conception of the aspects taken up for analysis. Some branches are now becoming more *atomistic* (micro approach), while others are integrating various atomistic results with a view to formulating a holistic background. Geography is one of the most promising scientific branches of the latter category. It has now developed "as a science concerned with synthesising the variable character of the earth surface as a whole or in parts".[1] The emergence of spatiotemporal approach has provided a new vitality to the subject.[2] Geographers provoke that "Geographic space is seen and evaluated in different ways at different times and in different cultures".[3] On the other hand "time" is not only conceived as merely an occurrence, rather as a channel of continuity and existence. With these two parameters, Geographers are able to provide guidelines for making rational and comprehensive development plans. Before discussing some of the basic concepts, it would be worthwhile to examine the

* Keynote Address by Professor R.L. Singh, Vice-Chancellor, Meerut University, Meerut-250001; and the Chairman, I.G.U.-WG "Transformation of Rural Habitat in Developing Countries".

scope and limits of geography as an integrating Science.

Geography as an Integrating Science

Geography is now being recognised as the Science dealing with the cause-effect relationship among the spatial biological, physical and human perspectives, that is why it searches to explain the world of reality and tangibility.[4] This growing thrust of scientific organisation leads everybody to feel himself "to be the geographer of his own space".[5] Such trends would provide a potential basis to geography for serving the society; as such the geographic achievements may be applied for the promotion of the quality of life and the fullness of human life.

Taking these view-points into consideration, the first question which strikes one's mind is, where does Geography stand? Does it have any specific place in the scheme of the sciences? If there is a distinct place for Geography, and if so to what extent is it able to present the real structures, processes, and variations prevailing over the earth's surface within the perspective of human life chances? Geographic approaches are lying in the four major sets of human knowledge in different forms, e.g. spatial analysis in the set of geosciences, physical geography in the set of physical sciences, biogeography in the set of biosciences and human geography in the set of humanities. Within these approaches four distinct themes have developed which attempt to cope with the phenomenological diversity arising from the efforts to describe and account for the pattern created by man on the earth's surface. These are locational-regional, structural, man-environmental, and cultural. These themes joined together form the system of holistic concern, where "the atoms (or elements) are seen as attributes which can be understood as parts of the whole (or set)" as opposed to "understanding the whole (or set) as a thing which is built up from the parts (atoms)".[6] It is in this continuity that developed geographers' core concept of *socio-spatio-structural environmentalism,* which exists everywhere in the field of human knowledge at different degrees and levels. Furthermore, to cope with this march, the nature of geographic methodologies has now changed from its traditional bias, with a continuous process of revitalisation of its dynamic approach, because the field is creating future knowledge along with the pull from the society for solutions to future problems.[7]

. Contemporary geography emphasised the essence of geography in the context of spatial relationship, where the term relationship is taken in all behavioural, physical and in abstract forms. Defining the frontiers of geographical research, Ackerman has raised five persistent problems:[8]

(i) The study of distributions in the abstract;
(ii) The perfecting of techniques of observation;
(iii) The investigation of processes involved in the distribution of a given class of phenomena;
(iv) The analysis of co-variation among processes as reflected in space relations; and
(v) The integration of data on the several processes with those of site, so as to show the full evolution – any pattern of earth-space relations in a given area or areas.

Ackerman's provoking for direct implications of Geography in the solution of societal problems has now been accepted as practice. Wells' dream for a "Science of future" has become reality with the advancement of Futurology. Geographers have also come up with their innovative aspects like 'spatial forecasting' and 'spatial prediction'. Important works have been presented for making future models or 'spatial organisation' suitable for a particular area.

Geography, indeed is academic in general, but is essentially an integral part of our socio-economic system; because everything lies over the earth's surface, so it responds to society's greater needs. There are two parts of our terrestrial space, i.e. man, and the human settlements where man gets stabilisation. Sometimes the question arises whether the geographers can play any role to make a better future? The answer will certainly be 'yes', because it explains the system of human life in space with emphasis on the dynamics of processes and structures.[9] This system may be termed in geographic sense as 'geochorosystem' which analyses the system of regulation,[10] control and automation with its own techniques so called as 'geocybernetics'. Such a perspective forces the geographers' thinking "to consider the constraints operating upon large man-environmental systems, and it greatly enlarges our geographic experiences, sensibility, and insight".[11]

Spatio-temporal View

The modern era of scientific and technological advancement impressed the geographers' mind sufficiently to innovate geography as a science of transformation with a view to formulating constructive guidelines and policy-frame to be applied for the promotion of human welfare, peace and love. It has rightly been observed that, "Our real need is for a new philosophy of man. In that philosophy, the human condition will be conceived as an evolving relationship between the contradictory forces of certainty and ambiguity, man and men, men and society, freedom and necessity".[12] This philosophy has made geography assume an important position in the field of behavioural and social sciences.[13] Gerasimov has rightly remarked that the main task of modern geography is to provide "the scientific backup for mankind's titanic efforts in the all-round rational use of natural resources, and purposeful protection and transformation (planning and change) of the environment for future development of social production and improvement of the living conditions of the population.[14] These themes and ideologies are expressed in spatial context in both forms – pattern and process – as such geographic approach necessarily considers man-environmental relationship as the integrative system; thus any region or area may be viewed as "complexes of interrelated distributional patterns, lines of movement, and spatial processes",[15] all involving change and transformation through time.

Within such approach in geographical science two basic concerns emerged dealing with space and territory, the fundamental property of the environment; as a consequence of interaction between the two there developed a 'geosystem', making geography play the role of an integrating science. It is notable that the concepts of space and territory are complementary to each other.

In spatio-temporal frame, any environment forms a distinctive complex system with many sub-systems and elements which advance the functional wholes providing inputs for other sub-systems from which they, in turn, receive inputs.[16] For full comprehension of the complex system, the unifying theme can be illustrated by three sets of concepts: (i) exchanges between systems and environments, (ii) energy and entropy, and (iii) diversity, stability, equilibrium and feedback. Within such conceptual frames, the cycling of resources may be taken as the transformation of matter within a system until it is again available for reuse; such transformatory processes depend on the use of energy as it flows through the various channels of the system. Thus any mature system can be taken as 'open' or 'closed' mainly on the basis of the trend and structure of energy flow and its channellisation. The organisational structure and the cyclic patterning in the system are the replica of the reorganisation of the feedback and diversity processes by which the system maintains its stability and equilibrium.

To analyse any geographical problem, which, in fact, is basically discriminatory in space and indeterminatory in time, the models can be used to analyse the structure on the one hand and to predict the future on the other. As such, the geographical models have two ways or so-called two distinctive procedures: 'one Kinomatic' with an aim to parameterise the forms/structures or the States of the process under consideration, and the other 'dynamic' describing the growth and evolution of those forms in time.[17] These two are complementary to each other; so, for a balanced analysis it becomes necessary to apply *Kinomaticodynamic* models which will deal with the cause and effect, the fundamental theme of spatio-temporal analysis.

Concept of Spatial Planning

As all growth processes and their consequence occur in space, the space must be treated as the fundamental base; as Secomski has remarked "This fact furnished sufficient reason for treating spatial planning and the future spatial structure as a point of departure or else as a foundation of the general economic planning.[18] Fundamentally, everywhere in any formation and progress, there are undecidable and invisible elements and happenings in time sequence (as propounded by Heisenberg in his theory of uncertainty). It is notable in this context that "the more accurately we measure one parameter the less accurately are we able to measure the other".[19] To describe and measure this factor the theoretical logicians innovated a language, ***metalanguage***, which analyses the trend with distinctive, definitive and directive nature.[20] Curry's forecasting map[21] dealing with space-time processes and Cliff and Ord's time-space processes[22] are perhaps the latest advancements made by geographers. With such basic conceptions, for any regional or territorial development the injective forces can be controlled, regulated and diffused from the nodal centres; it is primarily needed to prepare a spatial system of service centres of mono- and multi-characteristics at different levels, so that an integrated and balanced developmental strategy can be formulated and implemented at different levels.

The consciousness that the regional analysis provides an efficient base for the presentation of the real world situation rather than the normative, its analytical tools are applied for uncovering "the diverse organisational and institutional structures of society as they govern the behaviour and spatial distribution of population and activity"[23] Geographic space constitutes the natural environment in which man lives, acts, and rests.[24] Variation and disparateness are the fundamental characters of the natural environment. The imbalances found in the environmental system restrict the achievement of the ultimate goal, i.e. "the promotion of the quality of human life". It has been recently felt that balanced socio-economic-spatial organisational system can be achieved through spatial planning. The basic advocacy concerning planning postulates is related with the concept of economic development as well as with the settlement system. The basic aim towards the development of settlement system with its varying magnitudes and functional distinctiveness is reflected in the social and economic niches, which ultimately form an integration in regional development. It has been rightly remarked that "a settlement system is a spatial framework of the social and economic activities".[25]

The structure of spatial planning can be presented with a system which presents the interaction and feedback processes operating in the formation of spatiotemporal organisation. This is being governed by the locational behaviour of the two basic demands of man, social and economic, and their potentiality and level of utilisation. The appraisal and utilisation of resources are controlled by the combination of three attributes, plan, technology and interaction system.

In general, the concept of spatial planning has two-fold perspectives; firstly, it searches for the cause-effect relationship and its explanation, and further it gropes for practical understanding and its implication. Four types of approaches towards spatial planning may be distinguished:

(i) *Methodological*, providing methods and techniques as tools either indirectly for research carried on for planning, or directly in planning itself.

(ii) *Diagnostic*, assessing present conditions and the history which has led up to them.

(iii) *Normative* or *Predictive*, to determine what might occur under certain stated conditions or as the result of given actions.

(iv) *Therapeutic* or *Prescriptive* to prevent certain transformations, or recommend transformations to attain certain goals.[26]

The concept of spatial dynamics has been postulated for regional development process at three levels: homogeneity, interaction, and dynamic spatial processes. Spatial inequalities in development constitute a crucial problem in the territorial organisation of developing countries, which can be categorised into four types: (i) Inter-regional, (ii) Urban-rural, (iii), Inter-urban, and (iv) Intra-urban,[27] and measured in terms of various indicators. The spatial inequalities of the developing countries form their own distinction, especially with concentration of more population in the rural areas where traditional economy and culture form the base for operation. Diversified resource potentiality and imbalanced functional organisation with sporadic development at a few places without proper inter-connection and interactions are the basic characteristics of typical areas of such an environment.

In a planning strategy, the mode of production, social transformation, and the spatial organisation are important considerations. It is generally contended that differences between places result from the spatial ordering of a specific mode of production within a particular territory employing the polarisation system. Postulations have also been made towards decentralisation of production for balanced spatial development. Both the concepts are distinct – the former stressing upon more production, while the latter on better distribution. The identification of spatial reality in any development strategy involves the "dimension which is permanently engaged in its own readjustment under the influence of social and economic reality, but which at the same time exerts its own influences over that reality".[28] The processes of creation of space and of mode of production are inseparable, but the latter cannot be understood in the frame of the abstraction of the former.[29] It is, therefore, necessary to consider spatial reality at the first instance.

Development of spatial policy depends largely upon two considerations[30] to prepare spatial process models for application as a frame, and policy formulation for structural changes and transformation strategy. The identification and critical appraisal of these two steps are always to be considered as complementary to each other. Such a procedure makes it essential to consider man's behaviour especially proxemics – how the people in space accept the innovation and adjust accordingly? This dynamics is always understood in terms of time and space which are *functions of each other*, what anthropologist Hall has called as 'extension transference'. For any proper development strategy "full implication of extension as lesson and the extension as mirror have yet to be fully realised.[31]

Concept of Rural Habitat Transformation (RHT)

For the betterment of the future and maintenance of ecological balance in reference to rural habitat many theoretical approaches have been developed. Presently, regional planning is considered as strategy of established government machinery, but the prevailing conditions in the Developing World do not fit properly with the pragmatic response. This situation had led to the development of some more feasible and rational approaches, notable among which are agropolitan strategy, Integrated Rural Development (IRD), micropolitan development (MPD), basic minimum need programme (BMN), another development, and development from below. Within about past one decade, since the establishment of International Geographical Union's Commission/ Working Group on 'Rural Habitat Transformation in Developing Countries' (RHT-DC), under the chairmanship of Prof. R.L. Singh, attention has been drawn towards the formulation of the "Science of Habitat Transformation" in particular reference to the Developing World.

The concept of RHT is more comprehensive and based on 'need-value-ecodevelopment' programme with a view to promoting human welfare, while maintaining the ecological balance. Highlighting the critical role of space in organisation-system, "territory and function" has been conceived as the fundamental consideration in RHT programme. Like IRD, the RHT seeks to purvey multi-sector, multi-level and multi-section perspective of rural development; its /dynamic approach to transformation of rural habitat as a continuing process is more meaningful in real development context: an attribute which distinguishes it from IRD. The primary task of the RHT programme is to foster new initiatives with special reference to (a) conceptual framework including system analysis and model building, and (b) application of the results in preparing guidelines for various development programmes. Under RHT programme three fundamental dimensions have been identified: (i) *Socio-economic*, which includes human and infrastructural coordinates, (ii) *Environmental*, referring to ecological problems, resource appraisal and related aspects like health, and (iii) *Behavioural*, representing the proxemics and awareness. Our effort is mainly hinged to the structure of advancing knowledge about the scientific understanding of RHT as a continuing process in collaboration with international colleagues and institutions. Various issues need to be examined at different scales with an aim to maintain the basic objective, i.e. to bridge the gap between concept and reality, and policy and practice.

Among the commonly used and basic need strategy in reference to RHT, 'agropolitan approach' has been accepted widely. This approach has been purveyed within four dictums — agricultural, territorial, economic transaction, and administrative. The basic idea in this approach is 'to make possible a development that is geared to the satisfaction of evolving human needs'. This aim has also a temporal frame of growth and development, especially with the cultural revolution in the developing world, and also with the active participation of international institutions such as the UNO, UNESCO, IBRD, AID, ILO and the Club of Rome. Recognising the ideological gap between urbanistic and ruralistic approaches it would be more appreciable if a proper balance is maintained between these approaches within spatio-temporal frame. The application of the two concepts, 'eco-development' and 'system structure' (also involving cybernetics) will be a pre-requisite. The framework of RHT need be conceived as the strategic design of problem solving planning process. It is expected that the developing Science of Habitat Transformation will emerge as parallel to that of 'Futurology'.

Concept of Eco-development

The concept of 'eco-development' is used to describe a form of planned growth which is primarily concerned with the development of the locally available resources within the constraints of the local environment. The main object of this approach is to maximise the local capacities of the biosphere to support human life and assimilate wastes. This approach is very close to ecological and environmental planning referring to balance and harmony among the various enterprises, which man, for its own benefit, has superimposed on natural environments.[32] According to UNEP the concept of 'eco-development' has originally been designed "to support the effects of the people living in villages and other rural settlements to understand and utilise in their own development the basic natural resources and human skills available in their own environment".[33] With this advocacy this concept has been conceived as "an alternative strategy for development of rather general applicability and considerable promise not only for natural and rural areas, but also for some forms of urban settlements".

In reference to the institutional framework of eco-development, three basic-guidelines are necessary.

(a) Eco-development implies the creation of a horizontal authority able to transcend the sectorial approaches and to deal with various aspects of development, taking constantly into account the complementarity of various actions undertaken;

(b) Such an authority cannot be efficient without the participation of the population concerned in the working out of eco-development strategies. Such participation is necessary for the identification of concrete needs and of the productive potentialities of the eco-system, as well as for the organisation of the collective effort for its utilisation; and

(c) It is necessary to make sure that the participating populations for which the strategy is worked out are not deprived of its results by intermediaries who stand between local communities and the regional or national market.

It may be generalised that the eco-development approach rests on three ethical postulates: (a) development of human being in reference to well being rather than economic sense, (b) promotion of 'self-reliance' in context with decision-making, and (c) advancing a 'true symbiosis' between "man" and the "environment". To activate and regulate the harmony between development and environment both the "demand" and "supply" side of the equation must be acted upon.

Intergrated Frame of Development

It is the general consensus that IRD should include the whole range of activities for the betterment of rural environment. It has been stated that the transformation of rural environment is the fundamental need for the growth and development of the entire national economy. It is notable that the concept of IRD and RHT should be generally considered similar to each other, while RHT is also concerned with the ecological and environmental issues in addition to the integration of sectors and policies in space-time dimension. With a view to promoting the rural well being and the betterment of rural livelihood the integrated frame will include: (a) production components like agriculture and industry, (b) physical infrastructure coordinates such as roads, minor flood-embankments, drains, rural water supply, marketing, storage, housing, etc. and (c) social infrastructure attributes such as family planning, health, education, social works, etc.[34] To make integrated scheme more feasible several criteria have been given special attention for providing better services and promotions of development programmes: employment and minimum needs, choice of technology, reconciliation between sectors, development *vs.* distribution, viability, advance and consent, environmental balance, inter-sectoral independence, cost-priorities and phasing, and finally, implementation through coordination and financial control.[35]

Considering the environmental conditions of the developing world, attempts have been made to postulate the model of development. At the micro-level, having the main objective of 'human development' through process of 'structural transformation and mobilisation', a scheme of *another development* has also been discussed with a specific aim to 'delinking' and 're-linking' for the New Global Order.[36] Moreover, normatives have also been advanced pertaining to understanding the contradictions like rich *vs.* poor, landlord *vs.* landless, trader *vs.* labour, training of committed village cadres for initiating activities, building organisations of the poor, and strengthening viable situations. It has been strongly suggested that each criterion should be evaluated at first and ultimately integrated to provide for the strategic importance of balanced development.

The functional integration is to be considered in the light of at least four coordinates: (a) growth functions basically referring to marketing, trade and agricultural upliftment, (b) diffusion of innovation indicating the human choice and the level of adaptability for the newer ideas or means of development, (c) service provisioning attached to the spatial infilling and efficient use of resources for equity and balanced organisations and (d) coordination of development planning, which forms a base for efficient and dynamic form of capital investment, management and control.[37] It is notable that the prospects of planning procedures and the suggestions made will have to depend upon political decision-makers, especially the attitudes and the basic policy of the prevailing government.

Some Experience of Developed and Developing Countries

According to the estimates of the UNO, it is expected that in South

America rapid growth of urban population will presumably persist at least until the year AD 2000. But, by that period the percentage share of urban population would not exceed 40 per cent in East Asia, Africa and South Asia, while in the rest of the world this share is expected to be over 70 per cent as against the world average of 52 per cent. This tendency clearly indicates the predominance of rural economy even in the future in developing countries.

Several important aspects of development programmes in the developing countries are related with four basic questions[38]: firstly, the aspects of production forces related to organic farming, renewable energy sources, and appropriate technology; secondly, the issues of the social relations of production; thirdly, questions related to the role of trade, foreign investment, and aid; and finally, problems of over-population and population growth.

Most of the developing countries are primarily based on agriculture and primary production, therefore, application of 'agropolitan approach' will be more feasible. Such areas are distinguishable by "densely populated agrarian societies characterised by low profiles of social development, high rates of population increase, incipient based industrialisation, high external dependency, and rising indices of inequalities.[39]

In India more than three-fourths of total population (683 million in 1981) is living in the rural areas, and 65 per cent of the total workers are engaged in the agricultural sector, but they provide only 45 per cent of the total Gross Domestic Product (GDP).[40] In most parts of India several kinds of poverty traps can be observed, what Friedman called as the "Syndrome of collective poverty". It has been agreed that the basic social, economic and environmental goals of planning will be adequately achieved only with change in the spatial structure of activities. The basic problem in such perspective is associated with the existing non-viable settlements system in the nation; about 38 per cent of the total rural population lives in villages of less than 1000 population and another about 26 per cent in villages between 1000-2000. Such a system can hardly support for a balanced development of community facilities. No attempts have been made so far "to provide any link between public sector undertakings and the regional economy or to maximise multiplier effects, so that there has constantly been imbalanced or 'lopsided development". The approaches have lacked the basic concept that 'resources are space bound'. One of the pertinent problems always to be faced between regional and national objectives is the conflict in priority considerations, what Friedmann called 'Politics of Regional Planning'. In a country like India the planning strategy should aim at promoting the symbiosis of urban and rural processes for a balanced development. It is obvious that there should be at least one growth centre at five km distance with banking, production and marketing facilities so that the rural environment may be re-vitalised creating strong links between the rural and urban areas. The production and consideration of uncertain/negative and invisible effects also needs a significant consideration. If a region shows growing trend of GNP, it does not mean that there does not arise spatial inequality or disparity. It has been found that in the States having highest per capita income, like Goa (Rs. 2,426), Punjab (Rs. 2,278), Maharashtra (Rs. 1,903) and Gujarat (Rs. 1,623 in 1980), spatial inequality is higher in comparison to other States. It is notable that 'the Green Revolution has created new classes in the Punjab, both on the Indian and Pakistani side of the border, a rural elite on the one hand and landless labourers on the other'. Such instances prove that there developed a big gap between the upper and the lower classes which further causes even occurrences of violence.[41]

With an approach of 'growth for social justice, achievement of full employment in the rural areas in a period of ten years and removal of poverty', the Sixth Five-Year Plan of India laid special emphasis on basic needs of the people in lower income-groups, including health care services and facilities and other amenities. Population size and growth altogether form a hazardous problems. According to 1981 Census India's population reaches to 683 million. Sixth Plan has been revised accordingly. Moreover, the Planning Commission has also revised its population policy for long term objective with an estimate to reduce the net reproduction rate to one by 1995 so that the annual growth rate comes down to a manageable 1.2 per cent. The new population policy also aims at reducing the death rate and infant mortality to nine per thousand of the population and 60 or less per thousand live births a year.

In case of African continent, although most of the development programmes indicate internal change, they are based on external aid. The impact of westernisation is deeply rooted in the urban areas, while rural society is deprived and stratified. Such a condition has been responsible for widening the gap between urban and rural habitats. Whatever the approach of development may be applied, but for a radical socio-economic transformation there is a need for clear-cut policy and orientation with a view to rejecting the contradictory combination of internal disintegration and external integration. Most salient steps in this concern will be (a) an accelerated utilisation of the richest and priority-based resources, (b) better and maximum productive utilisation of human resources, (c) the planned dispersion of industry over the whole nation by means of bringing it closer to the resource locations and the consumption area, (d) the rational location and spatial development of agriculture with special attention towards regional specialisation, and (e) the improvement of inter-regional socio-economic links and the full use of all forms of transport and communication system.

Considering prevailing conditions of economy and society, the Developing World may be divided into three categories: Countries where available capital and skill are still insufficient to generate non-agricultural employment at the rate concomitant with the increase in the number of job seekers; Countries passing through transition period where the increase in demand for employment is of the same order of magnitude as the increase in non-agricultural job opportunities; and Countries where both the percentage and the absolute number employed in agriculture is decreasing.[42] It is notable that in the Developing World, rural development, from the economic points of view, no longer constituted a *prima facie* case. The basic need in such areas is related with the comprehensive plan having the notion of integration and societal consciousness. The agricultural and non-agricultural developments have to be strengthened side-by-side with a view to complementing each other and making for an integrated utilisation of the existing natural and human resources. Another programme should be associated with the dispersal of urbanisation and industrialisation. The implementation of such policies is basically related with the adaptive principles of parallelism in the technological level which determines the economic results of any rural development programme, compatibility in the exchange of specific inputs and outputs which refers to the governmental and structural relationship and their mutual associationship, and location of industries in accordance with the spatial coverage as well as basic requirements.

Some of the adaptive measures of rational policy-making in Developing World are promotion of appropriate and indigenous technology, maintenance of societal system in respect to environmental setting, equitable distribution, dispersal of agro-industrial activities, and finally promoting active people-participation.

Concluding Remarks

The diversion of geographic approach for the human welfare has inspired the workers to develop more scientific explanation and analysis for causal relationship between phenomena, so that the real world would certainly be presented. This thinking has provided a "great deal to contribute to the struggle to achieve a harmonious and stable set of man-environmental relations, and in particular way in which cultural adjustments will have to be made inside the ecological envelops". The well-being of the society as a spatially-variable condition should be the focal point of geographical inquiry, where all other branches will emerge and find the interrogative base of analysis, which will ultimately be helpful in promoting the quality of human life chances upto an optimal level.

With the scientific achievements, some reactionary "role played by technocratic conceptions of harmonising the interaction between society and nature" Laptov has also been noticed.[43] It is in this continuum that there is an urgent need for the science for Habitat Transformation, to be parallel and somewhere integrative with the *Futurology*. The above science stresses to promote the scientific and technological achievements to be fitted in accordance with the environmental setting and feasibility, and finally lead to promote the overall well being of the society. While avoiding the hierarchical and territorial disparities in thought and action, the *Science cf Habitat Transformation* would consider the whole world as one family to promote peace and love among the human community at global level.

NOTES

1. Singh, R.L., *Role of Geography as an Integrating Science*, Varanasi, NGSI, 1977, p. 1.
2. Singh, R.L. & Rana P.S., Singh, *Spatial Planning in Indian Perspective*, Varanasi, NGSI, 1978, pp. 7-9.
3. Sack, R.D., *Conception of Space in Social Thought*, London, Macmillan, 1980. p. 3.
4. Simmons, I., "To what extent is the geographer's world the Real World," *Area*, Vol. 8, no. 2, 1976, p. 82.
5. Sautter, G., "Some Thoughts on Geography in 1975," *Int. Soc. Sc. Jl.*, Vol. 27, no. 2, 1974, p. 232.

6. Atkin, R.H., "An Approach to structure in architectural and urban design, 1, Introduction and mathematical theory," *Environment and Planning*, B, Vol. 1, 1974, p. 55.
7. Garrison, William, *Future Knowledge*, F.K. Schaeffer memorial lecture no. 1, Deptt. of Geography, Univ. of Iowa, 1972, p. 3.
8. Ackerman, E.A., *Geography as the Fundamental Research Discipline,* Chicago, Deptt. of Geography, Res. Paper no. 53, 1958, p. 28.
9. Buttimer, Anne, "The Social Scientist and Changing Value," Paper read in Stockholm Conference on Future, 1974, p. 4.
10. Gokhman, V., "Theoretical Geography", *Soviet Geographical Studies*, No. 39, 1976, p. 77.
11. Gould, Peter, "Concerning a Geographic Education," in *Brighter Undergraduate*, eds. Lanegram, D.A., & Palm, R., 1978.
12. Olsson, Gunnar, "The Dialectics of Spatial Analysis," *Antipode*, Vol. 6, no. 3, 1974, p. 51.
13. Winsberry, M.D., "Social Sciences in the United States, 1948 to 1975," *Geogr. Review*, Vol. 67, No. 3, 1977, pp. 335-43.
14. Gerasimov, I., "The Scientific and Technological Revolution and Geographical Science," *Soviet Geographical Studies*, no. 39, 1976, p. 5.
15. Taaffe, E.J., *Geography*, Englewood Cliff, N.J., Prentice-Hall, 1970, p. 6.
16. Russwurm, L.H., "A System Approach to the Natural Environment," *Man's Natural Environment, A System Approach*, eds. L.H., Russwurm & E. Sommerville, North Scitvate, Duxbury Press, 1974, p. 3.
17. Thom, Rene, *Structural Stability and Morphogenesis*, Reading, Mass, W.A., Benjamin Inc, 1975, pp. 2-3.
18. Secomski, K., "The Contemporary Theory of Space Economy," in his ed. *Spatial Planning and Policy*, Warszawa, Polish Ac. Sc., 1974. p. 13.
19. Rastrigin, L., *This Chancy, Chancy, Chancy, World*, Moscow, Mir Pub., 1973. p. 26.
20. Beer, Stafford, *Platform for Change*, London, John, Wiley & Sons, 1975, pp. 7-8.
21. Curry, L., "Univariate spatial forecasting," *Econ. Geography*, Vol. 46, 1970, pp. 241-258.
22. Cliff, A.D. & Ord, J.K., "Model building and the analysis of spatial pattern in Human Geography," *Journal of Royal Stat. Society*, B, Vol. 37, 1975, pp. 297-348.
23. Isard, V., *Introduction to Regional Science*, Englewood Cliff, N.J., Prentice-Hall, 1975, p. 1.
24. Leszezycki, S., "The Factor of Space and its Role in Today's Economics," *Spatial Planning and Policy, Theoretical Foundations*, Warszawa, 1974, p. 30.
25. Koloszar, M., "Interrelationship between settlement systems and socio-economic development," in *National Settlement Strategies, East and West*, ed. Harry Swain, Luxemburg, IIASA, 1975, p. 68.
26. Kostrowicki, J., "A Key Concept: Spatial Organisation,: *Int. Soc. Sc. Jl.*, Vol. 27, no. 2, 1975, p. 329.
27. Slater, David, "Underdevelopment and spatial inequality, approaches to the problems of regional planning in the Third World," *Progress in Planning,* Vol. 4, no. 2, 1975, p. 101.
28. Rofman, A., *Designating Regions of Concentration of Economy in Central Argentina*, Buenos Aires, Ediciones STAP-Planteos, 1974, p. 18.

29. Vieille, Paul, "L'espace global du capitalismed organisation", *Espaces et Societies*, Vol. 12, 1974, p. 32.
30. Moellering, H. & W. Tobler, "Geographic Variances," *Geogr. Analysis*, Vol. 4, no. 1, 1972, p. 48.
31. Hall, E.T., *Beyond Culture*, New York, Anchor Books, 1977, p. 29.
32. Edingtons, J.M. & M.A., *Ecology and Environmental Planning*, London, Chapman & Hall, 1977, p. 1.
33. UNEP, *Speech of Executive Director*, Nairobi, 1973, p. 135.
34. Chang, H.T., "Integrated Rural Development," *Rural Development Technology*, eds. G. Singh, J.H. de Goede, Bangkok, Asian Inst. Tech. 1977, pp. 10-11.
35. Roy, P. & Patil, B.R., eds., *Manual for Block Level Planning*, New Delhi, McMillan and Company, 1977, pp. 86-87.
36. Wignaraja, Ponna, "From the village to the global order," *Development Dialogue*, no. 1, 1977, p. 41.
37. Singh, R.L. & Rana P.S. Singh, eds., *Rural Habitat Transformation in World Frontiers*, Varanasi, NGSI, Pub. 30, 1980, p. 257.
38. Yapa, Lakshman, "The Concept of the Basic Goods Theory," *Rural Habitat Transformation in World Frontiers*, eds. Singh & Singh, Varanasi, NGSI Pub. 30, 1980, pp. 259-80.
39. Friedmann, John & Weaver, C., *Territory and Function*, London, Edward Arnold, 1979, p. 194.
40. Singh, R.L. & Rana P.S. Singh, "Transformation Model for Rural Development," A geographical appraisal of spatial inequality," *Productivity*, Vol. 19, No. 4, 1979, p. 606.
41. Singh, R.L., "Transformation of Rural Habitat and Spatial Planning Strategy," *Regional Planning and Rural Development*, ed. L.R. Singh, G.B. Pant Soc. Sc. Institute, Allahabad, 1982.
42. Veitz, Raanan, *Integrated Rural Development, A Rehovot Approach*, Rehovot, SSC-PAD no. 28, 1979, pp. 8-9.
43. Laptov, I., *The World of Man in the World of Nature*, Moscow, Progress Publishers, 1979, p. 37.

4

GENESIS OF RURAL DEVELOPMENT

RAMLAL PARIKH*

GEOGRAPHY AND history are inter-dependent twins, and this coincidence reinforces the complimentarity and inter-relation of these two pioneering comprehensive areas of social studies. The pioneer of History, Herodotus, had underlined this when he described "Egypt as the gift of Nile". The important thing is the synthetic nature of both the subjects which has given them a significant place in the scheme of Social Sciences and more particularly in Social Studies. The artificial walls between Natural Sciences and Social Sciences have virtually disappeared in Geography. Further, the relevance of Geography to "Development" gets established from the synthetic approach of this discipline and its endeavour to rectify the ecological imbalances that have erupted in the process of Development — between Man and Environment and between Physical and Human Resources. It will hardly be of much value to deal with lifeless spatial aspects in isolation. Far more important is to study human activities on spatial basis. Geography's stress on total profile of man in a spatial context enables us to view human problems in totality.

Men have struggled hard to conquer Nature. The Conquest of Nature has been a driving force of all explorers and inventors. But after three centuries of hard struggle we find that nature is too abundant and cannot be wholly conquered. The leaders of scientific research are now veering round the view about futility of Conquest of Nature and have started pleading for Harmony with Nature. Geographers have a

* Address of the Chief Guest, the Vice-Chancellor of Gujarat Vidyapeeth Ahmedabad.

specific role in promoting Harmony with Nature by using their skills of discovering the inherent strength of nature for the benefit of man. The ecological imbalances created by illusory development processes have led to great ecological imbalances, the gravity of which is still not adequately comprehended. Geographers have this upper-most task of restoring the ecological balance, in all spheres of life, and like History they have to facilitate the ever-widening vistas of Geographical studies.

Before we examine the contents of rural development, we should first try to understand the meaning of the word 'rural'. Rural development does not take place by a mere process of making it non-urban. The word 'rural' here is not spatial alone but is to be viewed in a socio-cultural context. Rural society has its own social and cultural ethos. It requires a different type of approach — a kind of community approach. Rural development is not a negative concept but it is a positive concept of strengthening the rural society on the foundations of self-reliance in basic human needs and securing for them a social order which will protect them from exploitation from within the rural society as well as from outside the rural society. Thus the rural development is not a counter-part of urban development. Conceptually it is not only different from urban development but in many respects even contrary to its ethos. The very nature of Indian rural society divided into more than 5,75,000 villages with nearly 4,50,000 villages having a population of less than 1,000 persons, requires an indigenous strategy of development rather than extending some of the elitist strategies of development. There are hardly 5,000 villages with the optimal size of 10,000 people. The word 'rural' means totality of approach and not an aspectual approach. This lends a new meaning to the entire development process, which despite loud claims, is fragmented, aspectual and unintegrated. The rural society has a culture (which is not necessarily backward as is made out) of its own and Rural Development must take place within the framework of rural culture and not merely rural location. It therefore requires an approach and technique that are in tune with rural society. Rural society in India has a predominant place, as even after 30 years of planning it still constitutes 76.17 per cent of our population according to preliminary estimates of 1981 Census. 90 per cent of the rural population still depends on agriculture and of this 75 per cent depends on rain-fed agriculture. In terms of agricultural holdings 51 per cent of the holdings are still less than one hectare in size, and 90 per cent less than two hectares. In terms of gainful work 30 per cent are landless agricultural labourers and 51 per cent are cultivators many of whom do not have land of their own. It is not a question of transfer of technology or extending merely urban facilities to villages but it is a question of reconstructing rural society on certain solid foundations which will reinforce their capacity to withstand the onslaught of urbanisation which flourishes on exploitation of villages. Mahatma Gandhi had warned as far back as 1929 when he stated: "To serve our villages is to establish *Swaraj*, everything else is but an idle dream." He further stated in 1936 that "If the villages perish, India will perish too. It will be no more India. Her own mission in the world will get lost. Instead of village subserving cities, cities must subserve villages." In 1942 he further elaborated this idea by stating that "My idea of village Swaraj is that it is a complete republic and independent of its neighbours for its own vital wants and yet interdependent for many others in which dependence is a necessity." The word 'rural' therefore is not a spatial qualification or adjective to the word 'development' but is the very foundation of an alternative strategy of development.

What is the meaning of Development? It has became a catch-word of the contemporary age. It connotes different things to different people. The social scientists including even those who are votaries of 'Development' are seriously questioned about the meaning of the very term 'Development', particularly after the most disturbing revelations of the failure of the first UN Development Decade of 1960–70. It was in the beginning of the Second Development Decade that the claims of development have been seriously questioned. The assumption that the 'development develops' is itself challenged. Even the Chairman of the World Bank Mr. McNamara could not escape this debate on the meaning of the word 'Development' when in his annual speech at Belgrade instead of using the word 'Development' he used the word 'social indicators of progress'. Development is thus not necessarily identical with progress much less with equity. It is a disturbing discovery that development promotes disparity and in consequence far from removing poverty, enhances it. This is not a plea for abolishing developmental process as such, but it is a plea for radical change in the approach of development from mere growth to distribution and from vertical mobility to horizontal and concentric relationships advocated by Mahatma Gandhi. Modern nations are full of disparities among themselves as well as within themselves. The disparities are assuming great proportions and are denying fruits of development to under-privileged sections of the society, particularly in Third World or Southern World, though the Northern World also is not free from it. Enough data are coming forth to establish that even developed countries have a substantial Third World within them. The development process which has so far tended to remain a monopoly of economists will have to liberate itself from this domination and resort to a synthetic approach inherent in geographical studies. It is here that Geography becomes more relevant to development. The development process has to be more spatial and decentralised and should move upward from bottom rather than flow down from the top. It has been unquestionably established that the downward transmission of development process does not take place effectively. Development in this sense will not merely depend on physical infrastructure but will depend more on human resources which are abundant in all parts of the world and in all parts of our country. Development will have to learn from Geography, the science of horizontal distribution rather than vertical growth only and will have to reckon with impact of environment on the life of men and *vice versa.* Development must be a micro-process which may blow up into a macro-profile. The illusory nature of Development is born out by the non-distributive character of its process. Growth with social justice has largely remained a myth. The new thrust will have to be 'Development without disparities.'

Next comes the question of the appropriate place of rural development in our planning. While rural development is increasingly recognised as an important segment in our planning, it is not yet the foundation of our planning. What we — Social Scientists — need to work for, is to establish the inescapability of its foundational nature in any process of development. Peripheral programmes of Rural Development will not suffice. Despite several hundred crores of rupees spent in rural areas it is not helping the process of societal change that is so necessary for making development distributive. This is not a question of increasing allocations for various items of rural services. For example, on the question of increasing the input of energy for the spread of technology, it is necessary to find whether greater input beyond a "threshold point", far from promoting equitable development will instead not create more disparities. According to Illich, U.S.A. spends 45 per cent of the quantity of its energy on vehicles alone. It is estimated that an average American spends 1600 hundred hours per year in his or her car. America uses more fuel for its 250 million people than the 170 crore people of India and China put together.

When natural resources are used in such an extravagant manner, how can one dream of an egalitarian world order or an egalitarian national order? Use of energy beyond a threshold point accelerates disparities and benefits only a smaller percentage of population. This one illustration should establish the crux of the problem that energy crisis can be met not by increase in quantity of energy but by reduction in its consumption, particularly by developed sectors of the society. The programme of energy development should, therefore, be a programme of restraint in use of energy. Once the use of energy beyond a threshold point is restrained, there will not be an Energy Crisis and an equitable development process will emerge. Thus Illich's advocacy for Energy with Equity is very apt. He has further stressed "participatory democracy postulates low energy technology, and only participatory democracy creates the conditions for rational technology."

Rural Development thus is not only a different strategy but an alternative strategy to the prevalent development model which continues to foster new disparities rather than reducing them. This is a daylight revelation which must be recognised at least in India, with the ultimate aim of securing an International Economic Order based on this type of alternative strategy.

5

WHITHER DEVELOPMENT

B.C. PAREKH*

THE WHOLE language of the First World, Second World and Third World were invented about fourteen to fifteen years ago. Similarly, words like 'Development', 'Developing', 'Underdeveloped' were not in currency until about 1962. If you remember the colonial days, the so-called Third World was referred to as savage, primitive or a backward society. Africa was generally described as savage, India was generally described as backward. After Independence, when these countries began to develop, they protested very strongly at being described as backward or primitive.

And a new word was invented, used and understood for the first time in 1951, and that was the word 'Underdeveloped'. So there was already a contrast drawn between the Western countries who were regarded as developed and the Third World countries who were regarded as underdeveloped. I remember a very important conference addressed by Pandit Nehru many many years ago where he strongly took issue with the word ''undeveloped', because it implied that we were still savage. Several new countries took up the debate and the new word was substituted, 'underdeveloped'. Quite a lot of new countries began to ask what being underdeveloped meant, and underdeveloped with respect to what? Sometime in mid-sixties, and I remember the transition taking place, because it took place in front of my eyes in the United States, a new word was invented, and that was 'Developing Countries', and this is the word that has stuck with us for

* Presidential address of the Vice-Chancellor, The Maharaja Sayajirao University of Baroda.

all these years. I am surprised, the Seventies went away without any murmur or protest being raised and we have come to the end of the Eighties and I do not see any sign of somebody questioning words like developed or underdeveloped or developing or whatever, because behind this whole language lies an image, an image of what it is to be developed.

When we describe an individual as underdeveloped, we have in front of us a certain norm, a certain standard with respect to which we describe him as underdeveloped. When an Indian who is supposed to be about 5 feet 5 inches or 5 feet 6 inches high, is 4 feet 10 inches, we would say, he is underdeveloped or his height is below the norm. So, there is a certain standard and these standards vary from civilisation to civilisation and from society to society. Therefore, when a whole nation, and not only a whole nation, but a group of nations, is described as underdeveloped, or developing, one wants to ask what the underlying norm is.

What is it that constitutes development? Because it is this description and language which is extremely important because language is vehicle of thought and unless we are careful about the words we use, we are likely to be confused. Let us ask ourselves, what we have been doing with ourselves and what the West has been doing with us since 1947, when India was one of the first Third World countries to become free. This whole language that has been used by the western experts to describe us is an attempt to shape the whole of the Third World in the image of Europe or America. This has gone on since the days of colonialism. Undoubtedly, the Western world made a major breakthrough in human history around the 17th century in the aftermath of Renaissance and Reformation, when a whole new civilisation was developed within the womb of Christianity. That was the civilisation which generated rationalism, science, capitalism, a whole new attitude to rationality, the new spirit of individualism, the new spirit of economic development. All these things grew up simultaneously in the 17th century. And the whole of the Western world has developed in the direction dictated by this logic of development in the 17th century. And the path is already known. It is the path of capitalism, the path of industrial growth, the path of urbanisation, the path of developing individual personality, the path of developing the individual as an isolated atom. A whole new civilisation has developed in Europe and in the Western world since the 17th century. Urbanisation, industrialisation, developme-' growth of nuclear family, growth of individuality and so on. This is taken to be the norm. This is taken to be the standard or the ideal for the entire mankind. And it is this standard which is used in judging the rest of the world. If you judge India, for example, by this standard of urbanisation, industrialisation, breakdown of the joint family and development of the nuclear family, we are certainly undeveloped or underdeveloped.

But we are underdeveloped only if judged by this standard. If you substitute a different standard, for example, moral development, the richness of human relationship, the generosity of the spirit, the capacity to tolerate, to reconcile, to understand, then it would be hard to say that Western world is the most developed and we are the developing countries. We would instead say that we are developed because of the greatest contribution, in my view, India has made to human civilisation is the growth of the family. There is no civilisation in human history which has developed the institution of the family with all its complexities, with all its enormously rich vocabulary of describing the human relationship. There is no civilisation which has developed family to the degree that we have. The West has never known it. Africa has known tribes. China has no families in the way we have, given the sense of time. The Hindu or the Indian contribution to civilisation is the institution of the family. And we have developed complex social skills in organising the family. It is for this reason that we have tended to see the nation, the state in terms of the family. There is no other country in which, for example, the President of the Congress would be known as Rashtrapati, which is the husband of the nation; or that Gandhiji would have been known as Rashtrapita, which is the Father of the Nation, or that Pandit Nehru would have been described as *Chacha* Nehru; or where the meetings begin by addressing the audience as brothers and sisters. All these words — father, mother, *Chacha*, brother and sister — are family words and we project them on to the State, on to the nation because this is what we expect the nation to be, wittingly or unwittingly, intentionally or unintentionally. We have tried to conceive the new political system that we have developed in the image of the family. That is the most important institution upon which our civilisation is grounded. If you, therefore, take the development of the family as a standard by which to judge societies, then we would be regarded as most developed, and the West, where the family is in crisis, will be regarded as the least developed.

So, the point of the whole semantic exercise was to suggest that when we use words like developed and underdeveloped, we have all been brainwashed into a certain manner of thinking, a manner of thinking for which Westernisation constitutes development, and unless we challenge that manner of thinking we would not be able to sit back and ask ourselves where do we wish to go. To my view the Gandhian model of development is not the model that is available to us. The people of India have rejected it, rightly or wrongly. But at the same time, is Westernisation in the sense of individualisation, in the sense of developing nuclear families, in the sense of developing their extremely abstract and barren individual which is what characterises the Western civilisation? Is this the direction in which we wish to go? And as far as I can see, this debate in India almost ended in 1950 when we embarked upon the path of Westernisation, when Gandhian tradition of development kept lingering on in the background, on the peripheries, in the margin, making occasional noises, but never really making an impact. And since this is our imagination, our political imagination seems to have became restricted. Since our Independence, we have never asked ourselves if this uncritical emulation of Westernisation is the only answer.

What is in this Western model that we wish to imbibe? What is in our own tradition that we wish to retain, and how is it possible for us to absorb these elements of Westernisation which are absolutely crucial. There is hardly anyone here who is dressed in other than the Western manner. All our proceedings have been in English language. Our thinking has been in English language. And to think in English language is to think in an English way, because language is not merely a set of words, it is a mode of thought. Given the fact that so much of our cultural life, political life has been under Western influence, but given also the fact that we have retained the large part of our heritage, how to combine the two? How could we retain our heritage which we cherish and reject what we are ashamed of, whether it be our attitude to women or to animals or to nature or to the untouchables or to each other. How could we criticise our own civilisation, our own heritage in the light of the Western heritage to which we were subjected during the days of colonialism. And conversely, how could we criticise the Western heritage in the light of what is the best in our own? It is this debate that needs to be taken up. And I very well hope that this symposium will be an occasion which will trigger this debate which is long overdue.

PART II
LAND-USE AND RURAL DEVELOPMENT

6

LAND-USE COMPONENTS AND LAND-USE EFFICIENCY IN A PERIPHERAL REGION

RAM BALI SINGH AND SHRIKANT PANDEY

AN ANALYSIS of rural land-use becomes relatively more significant in the geographical study of a predominantly rural and overwhelmingly agricultural region. An analysis of spatial distribution and temporal changes in rural land-use components can help in explaining the spatial disharmonies in distribution and growth, in understanding the socio-geographical environment, and in estimating the development potentialities in agriculture and related activities. Land-use components, namely the area not available for cultivation, cultivable waste and net area sown as well as irrigated and double cropped area, have a tendency to show spatial and structural changes in response to variations in physico-economic and spatio-cultural factors and the process of rural development. In a tradition bound, small-scale subsistence agriculture, area not available for cultivation remains high because of several physical factors, and gradually declines with increasing population and developing agriculture but rises again with the development of physical infrastructures, social overheads and other agriculture based non-farm activities, services and institutions. Cultivable waste is influenced by the rural population pressure, marginal productivity and ecological awareness, level of technology as well as land and water management practices. Net area sown may decline during the process of agricultural and rural development. A relatively higher level of wages and agricultural productivity may sometime drive the marginal land out of cultivation; however, expansion in double cropped area with an increase in irrigational facilities will ensure a larger total cultivated area. Simultaneously, efficient land management and agricultural operations will result in a larger farm output and a higher level of rural income.

Objectives

The present study, based on village-level analysis of Pharenda Tahsil of Gorakhpur District, adjoining the Nepal border, attempts to identify some salient characteristics of spatial distribution of land-use components, to indicate certain trends in temporal variation of these components during a period of three decades and to evaluate landuse efficiency using the ranking coefficient method. An attempt has also been made to trace, if there exists, a gradual north-south gradient in the distribution of landuse components as well as landuse efficiency, as the study area is peripheral with reference to the nearest city (Gorakhpur), the general process of settlement and land occupancy (northward in the Saryupar Plain), the regional metropolitan core (Kanpur) and the national space economy.

The Study Area

Pharenda Tahsil (26°54′ - 27′N and 83°14′ - 83°42′E) covers an area of 1495.5 sq. km and has a population of 635,159 persons. It is almost a featureless alluvial plain having all the environmental characteristics of the *Tarai*. Most of it was forested during the early medieval period with the Tharus in the north and the Domkatars in some parts of the south. Later on the area was occupied by the Rajputs. During the Mughal period land cultivation was limited to certain pockets. During the 18th century the area suffered from invasions by the *Banjaras* from the north. Most of the area remained forested up to the middle of the 19th century. With the establishment of the British rule there started the process of expansion in the cultivated area through clearance of forest land. The British rulers allocated the land to individuals through auction at very low prices and sometimes through grants. This system of land-allocation resulted in the migration of population from the densely populated southern parts of the Saryupar plain to the relatively sparsely populated *Tarai* region in the north. A gradual decrease in density of population from south to north may be attributed to this process of colonisation from the south towards the north.

The study area has 636 villages including 30 uninhabited ones. The population density is about 425 persons per sq. km. Only 4.5 per cent of the total population is urban and only 19.0 per cent of the total population is literate. Workers constitute less than one-third (31.4%) of the total population. About 89.5 per cent of the total workers are engaged in agriculture (67.0% cultivators and 22.5% agricultural labourers) and about 93 per cent of the total cropped area is devoted to cereals. All indices indicate a traditional but now modernising agriculture in the study area.

Data and Methodology

The data related to five land-use components for the year 1941, 1945 and 1975 were obtained from the office records at Tahsil Headquarters (Anandnagar) and concerned village *Lekhpals*. Some of the data were obtained through personal survey of a few sample villages. In order to identify some salient characteristics of spatial distribution of landuse components and to indicate certain trends in their temporal variation, a series of maps were prepared showing the percentage distribution of the components in each and every village of the study area. In order to evaluate landuse efficiency the method of ranking coefficient has been used and to identify the north-south gradient the average ranks secured by each village with respect to five components have been shown over a map of the study area.

LAND-USE COMPONENTS

Area not Available for Cultivation

About 8.8 per cent (13,001 hectares) of the total area is occupied by this component. It constituted 8.2 per cent of the total area in 1945, declined to 6.7 per cent in 1955 but has been increasing since then because of an increase in area under human settlements as well as physical infrastructures and socio-economic institutions. Table 6.1 clearly indicates the distribution of this landuse component in 1945 and 1975.

Villages with lowest proportion of non-cultivable area are more frequent in the Tarai region in the north and flood affected *Khadar* in the south-west, while those with the highest proportion are relatively more frequent in the *Uparwar* and isolated elsewhere. A higher proportion of the non-cultivable area is usually associated with

TABLE 6.1: Area not Available for Cultivation

Category	Percentage of Total Area	Number of Villages		% of Total villages	
		1945	1975	1945	1975
I. Low	5.0	93	78	14.6	12.2
II. Medium Low	5.0–10	401	374	63.1	58.8
III. Medium	10.0–15.0	101	122	15.9	19.2
IV. Medium High	15.0–20.0	25	38	3.9	6.0
V. High	20.0	16	24	2.5	3.8
Total	–	636	636	100.0	100.0

socio-economic (human settlements, irrigation and transport developments, markets, educational institutions, hospitals and other socio-cultural institutions), and physical factors (waterlogging, flooding, changes in the courses of rivers). About 153 villages (24.1%) have witnessed a decline as a result of flood control measures, development of drainage as well as adoption of water enduring varieties of rice, while as many as 483 villages (75.9%) have recorded an increase in the area under this component. The latter is usually associated with a corresponding decrease in cultivable waste and cultivated area. Developments in physical infrastructures and socio-economic facilities, consolidation of holdings and growth of rural service centres and implementation of Food for Work and Rural Employment Programmes are some of the factors responsible for an increase in the area under this component. Village level distribution of amount of change in non-cultivable area shows a diffused pattern of spatial distribution. The number of villages having more than 15 per cent of their total area under this component has increased by 50 per cent during 1945–75, while the number of those villages having less than 5 per cent has declined by about 16 per cent during the same period. A sample study of six villages indicates that the land under settlements, irrigation channels and transport system as well as construction works related to socio-economic development has registered an increase varying from 45 per cent (Mogalha) to 125 per cent (Chautarwa), while area liable to waterlogging and annual inundations has registered a decline.

Cultivable Waste

This component indicates the potential for further expansion in the cultivated area. About 7.5 per cent (10,487 ha) of the total area comes under this category and it has shown a continuous decline since 1950 when it accounted for 26.1 per cent (38,407 ha) of the total area. The village level distribution of the cultivable waste is very uneven with its highest proportion (54.6%) being in village Sakasi (Middle *Uparwar*) and the lowest in village Chautarwa (south-eastern *Khadar*). As many as 415 villages lie below the regional mean (7.6%) and are usually associated with developments in flood control measures, drainage facilities and modernisation in agriculture. About 26 villages with very high proportion of cultivable waste (Mean + 3 S.D.) are scattered throughout the study area and, in most cases, are associated with forest lands as well as waterlogging and flood affected areas. Table 6.2 clearly indicates the distribution of villages in different categories of this component in 1941 and 1971.

TABLE 6.2: Cultivable Waste

Category	Definition	Number of villages		% of Total villages	
		1941	1971	1941	1971
Low	Less than Mean	412	415	64.7	65.2
Medium High	Mean + 1 S.D.	151	166	23.8	26.1
High	Mean + 2 S.D.	38	29	6.0	4.6
Very High	Mean + 3 S.D.	35	26	5.5	4.1
Total	—	636	636	100.0	100.0

The area under this landuse component has declined by 65.7 per cent during the past 33 years (1945–78). This has happened mainly because of increase in population, development of irrigational facilities, adoption of improved agricultural technology and introduction of new seeds. Like the villages with the highest proportion of cultivable waste, villages with high (Mean + 2 S.D.) and very high (Mean + 3 S.D.) change (decline) are usually situated near the forests and river channels. A few villages situated along the Rapti have witnessed an increase in the area of cultivable waste. However, the distributional pattern of this landuse component is much more complex and it is very hard to discern the gradient.

Net Sown Area

It is the most important component of rural landuse as regards the areal coverage and contribution to the regional product of the study area but fails to indicate the level of land management and overall rural development. More than two-thirds (101,365 ha: 68.8%) of the total area is under plough and it has witnessed a gradual increase since 1950 (72,988 ha: 44.6%). Village level distribution of the net sown area is very uneven. It varies from 43.8 per cent in village Khairat (N. Tarai) to 92.4 per cent in village Sondhi (Middle *Uparwar*). Most of the villages in very high category are concentrated in middle *Uparwar* as well as southeastern and southwestern *Khadar*. Villages of medium low category are located near the rivers, forestlands as well as in the Tarai region in the north. Net sown area accounted for 52.4 per cent of the total area in 1945 and increased to 68.8 per cent in 1975. It has increased by 31.9 per cent during the 1945–75 period. Table 6.3 shows the category-wise distribution of net sown area in 1945 and 1975.

TABLE 6.3 : Net Sown Area

Category	% of Total Area	Number of villages		% of Total villages	
		1945	1975	1945	1975
Very Low	30	10	Nil	1.6	Nil
Low	30–40	31	Nil	4.9	Nil
Medium Low	40–50	54	21	8.5	3.3
Medium High	50–60	131	44	20.6	6.9
High	60–70	213	126	33.5	19.8
Very High	80	11	148	1.7	23.3
Total	–	636	636	100.0	100.0

Villages with less than half of their total area under this component were 95 in 1945 but because of increasing population pressure, flood control, drainage provisions and improved seeds and irrigation facilities, the number of such villages declined to 21 in 1945. A sample study of six villages indicates that a higher increase in net sown area is mainly associated with higher density and higher growth rate of population. As many as one-third of the total villages have attained the optimum proportion and another one-third will reach the same in near future.

Most of the villages with relatively higher proportion of net area sown are concentrated in the southern *Khadar* and middle *Uparwar*. Nevertheless, the villages, recording more than 30 per cent increase in their net sown area, are scattered throughout the tahsil.

Irrigated Area

It has increased from 27.1 per cent of the total cultivated area in 1945 (11.2% wells and 15.9% others) to 48.1 per cent in 1975 (canals 1.9%, tubewells 10.8%, wells 9.9.% and others 25.5%) registering an increase of 141.4 per cent during a three decade period. There exist great variations in its village-level distribution. Its lowest proportion (1.67%) occurs in village Hardi Dali (Nautanwa Block). Table 6.4 shows the category-wise distribution of this landuse component in 1945 and 1975.

TABLE 6.4 : Irrigated Area

Category	% of Total Area	Number of villages		% of Total villages	
		1945	1975	1945	1975
Very Low	15	104	21	16.4	3.3
Low	15–30	212	79	33.3	12.4
Medium Low	30–45	151	129	23.7	20.3
Medium	45–60	116	158	18.2	24.8
Medium High	60–75	45	176	7.1	27.7
High	75–90	8	58	1.3	9.1
Very High	90	Nil	15	Nil	2.4
Total	–	636	636	100.0	100.0

The number of villages with less than 60 per cent irrigated area was 387 (60.8% of total villages) in 1975 while there were as many as 583 villages (91.7%) in this category in 1945. Most of the villages included in the very low, low and medium low category in 1975 are situated in the north while those of high and very high category are located in a northwest-southeast elongated belt in the middle of the study area.

There were 15 villages having more than 90 per cent of their total cultivated area under irrigation in 1975 while there was none in 1945. The number of villages of medium high, high and very high category increased from 53 in 1945 to 249 in 1975 while those of very low, low and medium low category decreased from 467 in 1945 to 229 in 1975. The irrigated area has more than doubled in as many as 198 villages and about 91 villages have registered an increase of more than 200 per cent. Villages of high (150-200%) and very high (200%) increase in irrigated area are found in the northern Tarai Region and the eastern half of the middle *Uparwar*. Villages with low (50-100%) and very low (50%) increase have their compact distribution in the southern *Khadar*. There are only a few isolated villages of this category in the north. Thus it is amply clear that there is a tendency towards obtaining a spatial balance in the distribution of the irrigated area. One may easily discern a gradual decline in the distribution of irrigated area as a proportion of total cropped area from south to north as well as an opposite trend (increasing northward) in the percentage increase in irrigated area.

Double Cropped Area

It is an indicator of cropping intensity and is highly correlated to the proportionate distribution of irrigated area. Double cropped area constitutes 41.6 per cent of the net sown area. Its village-level distribution is very uneven ranging between 11.2 per cent in village Pharenda in northern *Tarai* and 89.1 per cent in village Aligarh in southeastern *Khadar*. Table 6.5 shows the category-wise distribution of the double cropped area during the years 1945 and 1975.

TABLE 6.5: Double Cropped Area

Category	% of Net Sown Area	Number of villages		% of Total villages	
		1945	1975	1945	1975
Very Low	20	56	12	8.8	1.9
Low	20—30	217	28	34.2	4.4
Medium Low	30—40	221	149	34.7	23.4
Medium	40—50	107	260	16.8	40.9
High	50—60	30	125	4.7	19.7
Very High	60	5	62	0.8	9.7
Total	—	636	636	100.0	100.0

There were 187 villages having more than half of their net sown area cropped more than once in 1975 and most of these villages are concentrated in southeastern part between Peppeganj and Anandnagar and northwards upto Lakshmipur. None of the very high category villages is found in northern Tarai. Villages having less than 30 per cent of their net sown area under double cropping are usually located in the low lying Tarai region in the north subject to water-logging, and southwestern *Khadar* liable to annual inundation. As many as 358 villages (56.3% of the total) have their proportion of irrigated and double cropped area above the regional average.

The double cropped area constituted only about one-fifth (20.7%) of the net sown area in 1945 and has registered an increase of 163.9 per cent during 1945–75. Those villages with more than 160 per cent increase in double cropped area are mostly located in the Tarai region in the north and the *Uparwar* in the middle, while those with less than 80 per cent increase are scattered throughout the study region except in Lakshmipur block which is a transitional area lying in between the *Uparwar* in the south and the Tarai in the north. Increase in the area cropped more than once may be attributed to factors like development of means of irrigation, flood control, provision of drainage, adoption of improved and late varieties of wheat and early as well as water resistant varieties of paddy and increasing use of chemical fertilisers. A gradual decline from south to north in proportionate distribution of double cropped area is easily discernible. However, percentage increase in the double cropped area shows an opposite trend.

As agricultural landuse efficiency is primarily a function of latter three landuse components (net area sown as well as area irrigated and double cropped) indicating a potentially developing agriculture, it will, more or less, resemble the pattern of spatial distribution and temporal variation exhibited by these three components.

LAND-USE EFFICIENCY

In order to arrive at the coefficient of land-use efficiency village-level distribution of the landuse components — net sown area, uncultivable waste, cultivable waste as well as irrigated and double cropped areas — in 1975 have been used for ranking. The total ranks, thus secured by each village have been divided by five to obtain average rank for each village of the study area. The highest landuse efficiency is found in village Dharmdih in Campierganj Block while the lowest is recorded in village Kandhpur in Lakshmipur Block situated near the river Rohin. The village-level distribution of landuse efficiency is very uneven and may be arranged into five categories (Table 6.6).

Villages with the highest landuse efficiency show a very scattered distribution; however, nearly all of them are concentrated in a narrow area extending in a northwest-southeast direction in the middle of the study area in Campierganj and Pharenda Blocks. Villages with high efficiency show a compact distribution in the same elongated area and are scattered elsewhere with a minor concentration in the northeastern Tarai region. Such villages are also frequent in Brijemanganj and

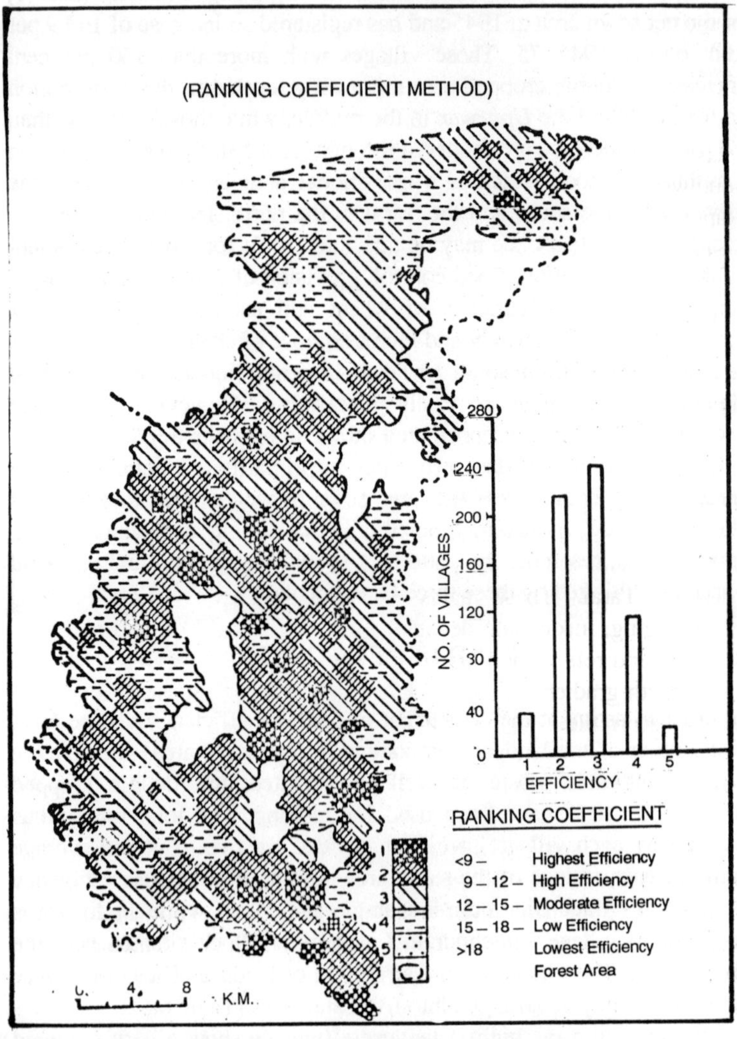

FIG 6.1: Pharenda Tahsil : Land-use Efficiency, 1975

TABLE 6.6: Land-use Efficiency, 1975

Category	Ranking Coefficients	Number of villages	% of Total villages
Highest	9	37	5.8
High	9—12	216	34.0
Moderate	12—15	241	37.9
Low	15—18	116	18.2
Lowest	18	26	4.1

Lakshmipur Blocks. Land-use efficiency gradually decreases towards southwest in flood prone *Khadar* lands of Ghughli, Rapti, and Burhi Rapti and towards northeast in low-lying Tarai region earlier susceptible to waterlogging and a laggard in the adoption of improved agricultural and land management practices. Villages having the highest and high landuse efficiency are usually associated with high population density, fertile soil, availability of physical infrastructure, nearness to service centres, cultivation of cash crops and improved agricultural practices.

Though the villages having moderate landuse intensity are found throughout the study area except in the southeastern part, however, such villages are more frequent in Lakshmipur Block and northern portion of Anandnagar Block. Villages having low and the lowest landuse efficiency are mostly found in the southwestern Khadar and northern Tarai. These areas suffer from annual flooding and waterlogging, inadequate development of transportation and irrigation facilities and relatively a larger distance to the rural service centres. A south-north gradient in the spatial distribution of landuse efficiency is easily discernible.

Recommendation

To improve the quality of life in these low landuse efficiency villages high priority should be given to removing the above mentioned constraints. The most urgent need is to check flooding and remove waterlogging. This should be followed by provision of transportation facility and rural service centres.

7

STRATIFICATION OF CROP COMBINATION REGIONS
A Parametric Prelude to Integrated Rural Development

DINESH C. PANDEY

THOUGH AGRICULTURE has conventionally been practised since time immemorial, recent advancements in diverse fields of knowledge have brought about drastic changes, and their repercussions which sprawl in almost all activities of man, encompass the primary activity of agriculture also. Needless to mention that pure sciences and various other allied disciplines have contributed much towards agriculture which has engendered that condition where planning redeploys technical advancement to channellise the scientific redundancy.

Integrated rural development is one of the major policies being intensively studied and implemented especially in cases of developing agrarian economies having a technical flummox. Within this, falls the Crop Combination Region (CCR) technique which envisages certain micro-spatial variables to form a contributing parameter to the integrated area development.

The variables of the CCR technique include the gross cultivated land, the number of crops, crop producing land and a pre-ascertained normative variable. These variables are synthesised to attain the value of crop combination (CCV), which reflects the optima for the number of crops which are sown and harvested in a particular region.

Study Region

The study region – Tahsil Kashipur (29° 6′ 36" to 29° 35′ 26" N and 78° 43′ 12" to 79° 18′ 58" E) is situated on the south-western border of district Nainital in the Kumaun Commissionary which lies at the north-eastern end of the State of Uttar Pradesh (Fig. 7.1)

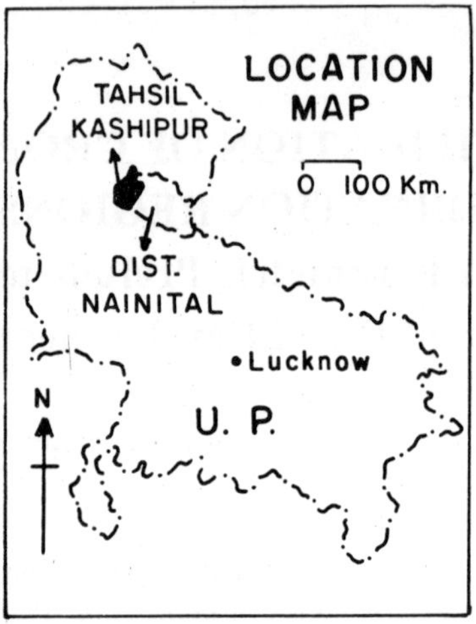

FIG 7.1: Location Map of Tahsil Kashipur

Operational Step

The aim of this paper in the identification of crop combination regions and their spatial distribution and intensity. From among the various techniques of delineation of crop combination regions, the author has adopted the method of Weaver with some modifications.[1] To obviate the necessity of the bulk of cumbersome calculations in Weaver's method, the author has logrithmically expanded his formula in the following manner:

$$\delta = \left\{\frac{1}{2}\log(2\log d - \overline{\log n})\right\}^{-1}$$

where δ = value of crop combination,

d = differences in actual crop percentage in theoretical curve, and

n = the number of crops in a given combination.

For the purpose of the study the percentages of cropped land occupied by different crops in Rabi and Kharif to the gross cultivated area have been calculated. Crops occupying less than one per cent of gross cultivated land have been ignored for being negligible in comparison to the dominant crops in the locality. Hence for the delineation of the crop combination regions nine crops have been taken into account – *Urad*, *Masoor*, Gram, *Arhar* and Peas, considered combined as pulses. Thus, the total five crops selected and designated symbolically are:

R = Rice, W = Wheat, S = Sugarcane, P = Pulses and M = Maize.

Calculations for all crop combinations for the total 48 clusters have lead to determination of the index value of crop combinations (Table 7.1)

Discussion

In the study region as a whole 43.08, 41.91, 26.07, 10.62 and 9.12 per cent of the gross cultivated land had rice, wheat, sugarcane, pulses and maize, respectively (Table 7.2).

The index value of monoculture, bi-crop, tri-crop, tetra-crop and penta-crop combinations have been calculated for the entire region. The deduced results show that the region falls under bi-crop combination category – its index value being 7.53 for the entire study region.

After making an analysis of the crop combinations by micro units (i.e. clusters), it has been found that 77.08 per cent clusters fall in the bi-crop combination, whereas tri-crop, tetra-crop and penta-crop combinations occupy 14.58, 4.17 and 4.17 per cent of clusters respectively. Out of the total bi-crop combination clusters 83.78 per cent is dominated by RW, and this area is equal to 64.58 per cent of the total micro-units of the region (Fig. 7.2).

In the bi-crop combination regions, 65.57 per cent land is occupied by RW, WR and WS, in which 53.57 per cent land alone comes under RW. Tri-crop, tetra-crop and penta-crop combinations possess 14.49, 10.59 and 9.35 per cent land respectively to gross cropped area in the region. Keeping in view the spatial pattern of various crop associations, the region has been categorised into six major agricultural zones or regions (Table 7.3 and Fig. 7.3).

TABLE 7.1: Clusterwise Crop Combination in Tahsil Kashipur

Name of the clusters	Index value of crop combination	Crop combination category	Crops in combination
1. Kashipur	8.70	Bi	RW
2. Ramnagar	3.42	Bi	RW
3. Jaspur	5.63	Tri	WRS
4. Barkhera Pandey	7.17	Tetra	SWRP
5. Megha Wala	6.69	Penta	RWSPM
6. Garhi Negi	7.96	Bi	RW
7. Ujjain	11.99	Bi	RW
8. Maldhan Chaur	6.37	Tri	WRS
9. Dhakia Kalan	1.98	Bi	RW
10. Mahua Khera Ganj	6.99	Bi	RW
11. Pirmadara	7.67	Tri	RWS
12. Raipur Khurd	6.28	Bi	RW
13. Lalpur	8.43	Bi	WS
14. Barhyonwala	7.98	Bi	WR
15. Dharampur	0.48	Bi	RW
16. Pratappur	7.65	Tri	WRS
17. Kundesari	2.31	Bi	RW
18. Chilkia	3.22	Bi	RW
19. Karanpur (R)*	0.48	Bi	RW
20. Patrampur	5.43	Tetra	RSWM
21. Sanwaldeh West	8.36	Bi	RW
22. Lakhanpur	1.63	Bi	RW
23. Chhoi	8.55	Tri	RWS
24. Dhanauri Patti	6.66	Penta	SPRWM
25. Mahua Dabra	0.71	Bi	RW
26. Nadehi	2.10	Bi	RW
27. Haripura	14.54	Bi	WR
28. Gopipura	0.24	Bi	RW
29. Bharatpur	3.46	Bi	RW
30. Jagatpur	5.41	Bi	RW
31. Surajpur	12.00	Bi	WR
32. Rahmupur	6.09	Bi	RW
33. Kaniyal	6.48	Bi	RW
34. Puranpur	0.40	Bi	RW
35. Jaspur Khurd	0.93	Bi	RW
36. Pat Kot	1.80	Bi	RW
37. Shivlalpur	7.47	Tri	RWS
38. Bansa Khera Kalam	1.77	Bi	RW
39. Gangapur Gosain	22.86	Bi	RW
40. Semal Khalia	6.62	Bi	WR
41. Kunda	8.03	Bi	RW
42. Jogipura	0.26	Bi	RW
43. Rampura	12.07	Bi	RW
44. Karanpur (K)*	14.43	Bi	RW
45. Talabpur	6.66	Bi	RW
46. Lakshmipur Lachchhi	6.60	Tri	RWS
47. Ganeshpur	3.46	Bi	WS
48. Kua Khera	15.91	Bi	RW
Tahsil Kashipur	8.59	Bi	RW

* Ramnagar Block. * K = Kashipur Block.

The Northern Rice-Wheat Region

The region consists of foothills and Bhabar area of Tahsil Kashipur and falls between 400 and 1000 metres of altitude. Patkot and Ramnagar are the only two clusters in the region. The agricultural patches in Ramnagar cluster always lie below 500 metres of altitude, whereas in Patkot such patches are found generally between 600 and 900 metres. A greater part of this region is covered with forest land. Rice and wheat are the dominant crops of this region. The per hectare production of crops is low but the areal strength of crops is considerably higher than normal, with the result that the agricultural efficiency falls in high degree of efficiency category.[2] This occupies 9.99 per cent crop producing land of the gross cultivated area of the study region.

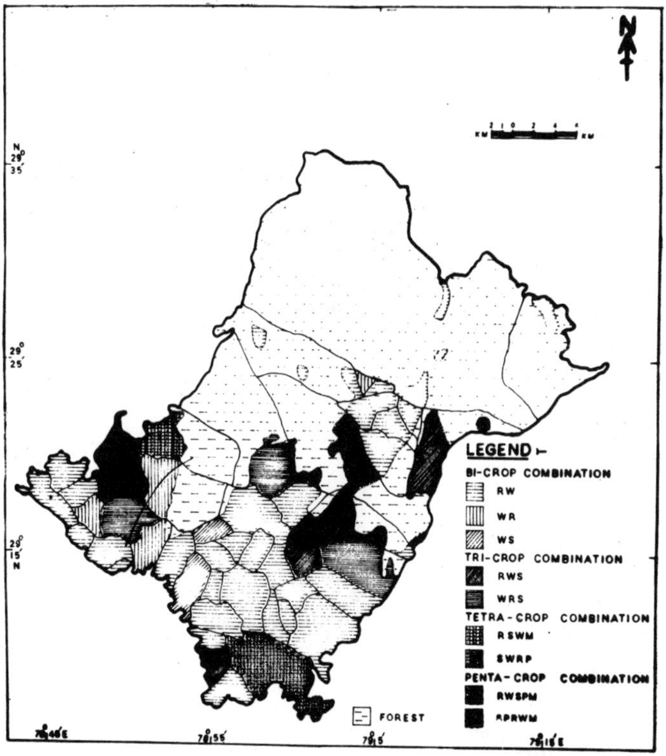

FIG 7.2: **Tahsil Kashipur** : Crop Combination Regions

TABLE 7.2 The Crop Combination of the Tahsil Kashipur as a Whole

Crop combination	Per cent of crop land occupied	Per cent of theoretical base	Differences squared	Crop combination index
A. Monoculture				
(i) R	43.08	100.00	3239.88	56.92
B. Bi-crop combination				
(i) R	43.08	50.00	47.88	
(ii) W	41.91	50.00	65.44	7.53
C. Tri-crop combination				
(i) R	43.08	33.33	95.06	
(ii) W	41.91	33.33	73.61	8.59
(iii) S	26.07	33.33	52.70	
D. Tetra-crop combination				
(i) R	43.08	25.00	326.88	
(ii) W	41.91	25.00	285.94	14.32
(iii) S	26.07	25.00	1.14	
(iv) P	10.62	25.00	206.78	
E. Penta-crop combination				
(i) R	43.08	20.00	532.68	
(ii) W	41.91	20.00	480.04	
(iii) S	26.07	20.00	36.84	15.85
(iv) P	10.62	20.00	87.98	
(v) M	9.12	20.00	118.37	

TABLE 7.3: Crop Producing Land Under Various Crop Combinations

Crop combination	No. of clusters	Per cent of clusters	Cultivated land (in ha)	Per cent of cultivated land
A. Monoculture	—	—	—	—
B. Bi-crop combination	37	77.08	28180	65.57
(i) RW	31	83.78	23024	81.70
(ii) WR	4	10.81	3826	13.58
(iii) WS	2	5.41	1330	4.72
C. Tri-crop combination	7	14.58	6226	14.49
(i) RWS	4	57.14	3492	56.09
(ii) WRS	3	42.46	2734	43.91
D. Tetra-crop combination	2	4.17	4553	10.59
(i) RSWM	1	50.00	823	18.08
(ii) SWRP	1	50.00	3730	81.92
E. Penta-crop combination	2	4.17	4017	9.35
(i) RWSPM	1	50.00	2562	63.78
(ii) SPRWM	1	50.00	1455	36.22
Tahsil Kashipur	48	100.00	42976	100.00

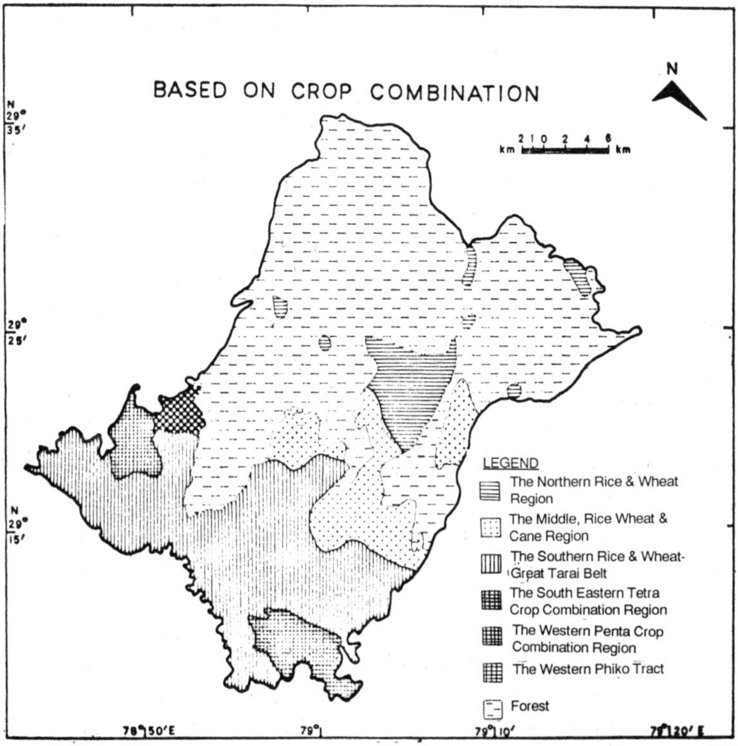

FIG 7.3: Tahsil Kashipur : Agricultural Zones

The Middle Rice-Wheat Sugarcane Region

This region occupies 14.70 per cent of the gross cultivated area. Use of advanced agricultural methods and implements is practised in this region. Kosi and Dabka rivers in the east and Dhela and Phika rivers in the west fulfil the irrigation requirements. The whole region falls below 400 metres of absolute relief. The relative relief is about 100 to 120 metres. Rice, wheat and sugarcane are the three main crops of this region.

The Southern Rice-Wheat Great Tarai Belt

This Tarai belt covers 58.75 per cent of Gross Cultivated Area (GCA) of the area under study. The belt has been designated as the "Rice and

Wheat Basket" of Kumaun. The soil is most fertile being alluvial in nature. The agricultural revolution in terms of mechanised agricultural practices and widespread extension of arable land can be seen in this belt. Due to higher productivity of crops, this region produces rice and wheat in surplus. Phika, Dhela and Kosi rivers are the main sources of irrigation. In this region, Tumaria Dam is one of the best examples of water management, which irrigates the north-western part of this region. In view of the above this is one of the most important agricultural zones or regions not only of Tahsil Kashipur but also of the entire Nainital District as well as Kumaun as a whole. The altitudinal variation falls in between 220 and 300 metres from MSL.

The South-eastern Tetra-Crop Combination Region

This region includes only a small portion of agricultural land of the area under investigation, which is about 5.96 per cent of the GCA of Tahsil Kashipur. Sugarcane, rice, wheat and pulses are the dominant crops of this region. Maize is the next important crop, which helps in increasing the agricultural efficiency of the region.

The Western Tetra-Crop Combination Region

This tetra-crop combination region occupies the smallest part of net cultivated area of the study region. A greater part of this region comes under forest cover. Canals originating from the Phika river are the main source of irrigation. Average elevation of this region is below 280 metres above MSL. Rice, sugarcane, wheat, maize and pulses occupy 29.16, 27.95, 24.42, 15.43 and 10.57 per cent of the GCA of this region respectively.

The Western Phika Tract

This penta-crop combination region is spread over only 8.68 hectares of agricultural land of the region. This region falls between 260 and 280 metres of height above MSL and lies below the average of crop and agricultural efficiency of the region.

The Phika river divides the region into two parts. The western part is covered by forest while the eastern part is mostly under cultivation, so that the level of production is not up to expectation. Rice, wheat, sugarcane, pulses and maize occupy 21.00, 19.68, 18.22, 13.27 and

7.19 per cent of agricultural land of this particular region. Barley is sown in a negligible area (i.e. 0.31 per cent of the gross cultivated area of this region).

Crop Combination Regions and Integrated Rural Development

A greater part of population of the area under study is directly dependent on agriculture and the region comes under the vegetation civilisation (i.e. agriculture moulds the economy and distribution of population of the region). The development of the region is a result of its balanced agricultural economy. A study of crops in their regional perspective must take into consideration the combinational analysis to understand the absolute and relative position of each and every crop. A particular combination of crops and relative importance of each crop in an area is helpful in interpreting and understanding some aspects of social and economic geography of the region.[3]

The concepts of agricultural efficiency and crop combination regions are interrelated. For example, the study region has been divided into five main crop combination regions and the bi-crop combination region occupies 65.57 per cent of the crop producing land of the region. This means that a greater part of the region is very much suitable for rice and wheat (RW) cultivation. The areas where multi-crop combinations are preferred by the people are not quite suitable for one or two crops and that the efficiency of the several crops will definitely be lower. Just the reverse is the case of mono-culture and bi-crop combinations. In the study area, the efficiency and productivity of the crops obviously govern the distribution of crop combination regions.[4] Thus, we state that as the index of crop combination increases, the degree of efficiency of crops decreases. The rice and wheat dominated southern Tarai belt of the area, has a very high degree of efficiency in rice and wheat cultivation in comparison to the areas of multi-crop combinations. Thus, for a concomitant and balanced rural development, the efficiency of crops must be enhanced which can be possible only by planting one or two crops (e.g., rice and wheat) in a region by developing adequate facilities of irrigation, pesticides, high yielding varieties (HYV) of seeds and fertiliser etc.

The basic and fundamental aim of integrated rural development (IRD) is the optimum spatial distribution of infrastructure (e.g., agricultural, industrial, basic and socio-economic infrastructure) for the optimum utilisation of resources to stimulate economic activities for the fulfilment of the social needs of the people. Crop combinations affect agricultural efficiency and agricultural efficiency directly influences social and economic way of living of the people. Hence, with the help of crop combination and crop efficiency indices, more rational the rural development planning is possible .

Conclusion

It may thus be concluded that the region is highly fertile, rich in other agricultural inputs and more suitable for paddy and wheat cultivation. Thus, in view of area planning, especially for rural development, the farmers of the region should prefer those crops which can be optimally produced in keeping with the physical as well as cultural factors. It has been observed and felt that in the monoculture and bi-crop combination regions, the technological revolution in agricultural activities can be more beneficial to increase the level of production in comparison to regions having tri-crop, tetra-crop and penta-crop combination or having diversified crop structure.

The above discussions make it very clear that a comparatively advanced agricultural level has been attained in the southern rice-wheat growing great *Tarai* belt. Due to more suitable physical characteristics, and due to being located in the sub-montane tract of Uttar Pradesh, and also due to more advanced cultural practices in agriculture, this region has become one of the best examples of rural development in U.P. in general and the Himalaya and the Tarai-Bhabar region in particular.

NOTES

1. Weaver, J.C., "Crop combination regions in the Middle West", *Geog. Rev.*, Vol. XLIV, 1954, pp. 175-200.
2. Pandey, D.C., *Tahsil Kashipur: A Study in Integrated Area Development*, (Ph. D. Thesis), Kumaun University, 1981, pp. 362-376.
3. Mohammed, A., *Studies in Agricultural Geography*, Rajesh Publications, New Delhi, 1978, p. 82.
4. Singh, O.P., and Pandey, D.C., "Spatial Variations in Agricultural Efficiency with Special Reference to Integrated Rural Development," *Himalayan Research Studies: Problem Oriented Papers*, pp. 82-88.

8

IMPACT OF HYBRID COTTON ON THE CROPPING PATTERN OF AMOD TALUKA, District Bharuch (Gujarat)

S.M.H. RIZVI AND D.A. BHATT

IN THIS paper an attempt has been made to investigate the impact of hybrid strains of cotton on the cropping pattern of Amod Taluka (21′ 47′–22′ 2′ N lat. and 72° 45′–73° O′ E long.) of District Bharuch in the State of Gujarat. It is observed that against the assumptions and aspirations of advantages, these strains have largely proved disadvantageous to the cotton culture of the whole taluka in general and of its western segment in particular. Thus it has adversely affected the process of rural development.

Physical Background

In this taluka, there are no pronounced terrain conditions. It is a gently east to west rolling alluvial plain. Among the major accountable relief features is River Dhadhar flowing from east to west, marking the northern limit of the taluka, and the estuary at its confluence with the Gulf of Khambhat. The back-waters of the Gulf spreading through the creeks have turned the western coastal fringe into a vast marshland, useless for agricultural purposes. The creeks or the backwater channels are the back and forth carriers of tidal and rain water from and to the Gulf.

Climatically, it comes under the influence of the South-West Monsoon. Rains come during the period mid-June to September. Like other parts of the country it faces the excess and scarcity situations of rain. Average amount of rain received between 1970 and 1979 was 823 mm with an average of 43 rainy days. Marine influence normally keeps the percentage of humidity relatively higher than the adjoining areas. Thus the amount of rainfall, the average number of rainy days, and the percentage of humidity are the most ideal climatic conditions for cotton. Table 8.1 gives the rainfall received between 1970 and 1979 and the number of rainy days.

TABLE 8.1: **Amount of Rainfall and Number of Rainy Days**

	1970	1971	1972	1973	1974	1975	1976	1977	1978	1979	Av.
Rainfall mm	1124	1009	507	902	328	991	1110	706	728	827	823
No. of rainy days.	49	34	29	47	35	56	58	49	37	38	43

Source: Taluka Office, Amod.

Table 8.1 shows that the rainfall situation in the given decade, except in 1974, was quite ideal for any crop, and more ideal for cotton.

Usually small gaps occur between diurnal and nocturnal range of temperatures. The average maxima and minima for a decade (1970–79) were 36°C and 28°C respectively. The sun is mostly bright, frost possibilities being quite rare. Summers are hot and humid, and winters mostly warm and dry.

Edaphically, Amod is the part of the black cotton soil belt of South Gujarat (locally known as "Kanam Vibhag", i.e. the region of black soils). Barring a few patches of alluvial loamy soils in the riparian margins, the entire Taluka enjoys deep, fertile black soils rich in humus and minerals, ideal for cotton crop. But the frailty of this soil is marked with its least tolerance to heavy showers and long-drawn dry spells. The most ideal rain conditions tolerable to this soil is light showers of 50-60 mm at regular intervals of 10-15 days, during the growing season. Under these rain conditions a successful crop is ensured.

Thus, the combined role of topography, climate and soil make the whole of black soil region, in general, and the Amod Taluka in particular, an important and outstanding segment of the cotton belt of Gujarat.

Regional Diversity

In spite of the superficial homogeneity of topography, climate and soil, the subterranean heterogeneity does exist. The underground water resources show marked differences between the western and the eastern segments. The western segment, being under the direct influence of the Gulf contains brackish sub-surface water reservoir in contrast to the sweet water of its eastern counterpart. The underground acquifer has been the only source of irrigation in the Taluka. Since, no natural dividing line between the two types of sub-surface acquifers exists any clear cut line of demarcation cannot be drawn. However, the Bharuch-Jambusar narrow-gauge railway line roughly serves this purpose, as it passes, somewhat along the line of divergence between them. However, it may not be a true geographical line of demarcation.[1]

The heterogeneity of the sub-surface acquifer, thus divides the taluka into two regions. The eastern part with its 35 villages is taken as Region I and the western part with 17 villages as Region II (Fig. 8.1). Region I with its 24,855.12 ha (61,417 acres) covers 56 per cent, and Region II with 19,760.53 ha (48,840 acres) covers 44 per cent of the taluka. In spite of the superficial homogeneity, the diversity of the subterranean acquifer has caused wide differences in the pattern of agricultural economy of the two regions.

Agriculture has been the mainstay of the people of Amod including those living in the town. The edaphic and climatic conditions are favourable for various crops, but cotton and wheat have been the overwhelmingly dominant crops, and the 'Kanam' region owes its fame because these two crops.

Cropping Pattern – A Retrospect

Peeping into the past through two to three decades or so, it is seen that the lion's share of the cropped land has always been taken by cotton, followed by cereals (wheat, *jowar* and *bajra*) pulses, and fodder. Table 8.2 shows the distribution of crops in Amod in 1955-56 and 1958-59.

Table 8.2 shows the significance of cotton at both points of time. Out of 93881 acres of NSA in 1955-56 cotton held 65.67 per cent, cereals 14.81 per cent, all pulses 11.76 per cent, and fodder 7.67 per cent. *Tur* among the pulses, tobacco and spices were quite insignificant. In 1958-59, cotton decreased by 3.61 per cent, and fodder by 0.17 per cent but, cereals and pulses increased by 0.84, and 3.04 per cent respectively. In spite of this minor decline, cotton dominates the landscape holding the largest share of the cropped area. It thus reveals the status of Amod taluka as the outstanding cotton cultivating area;

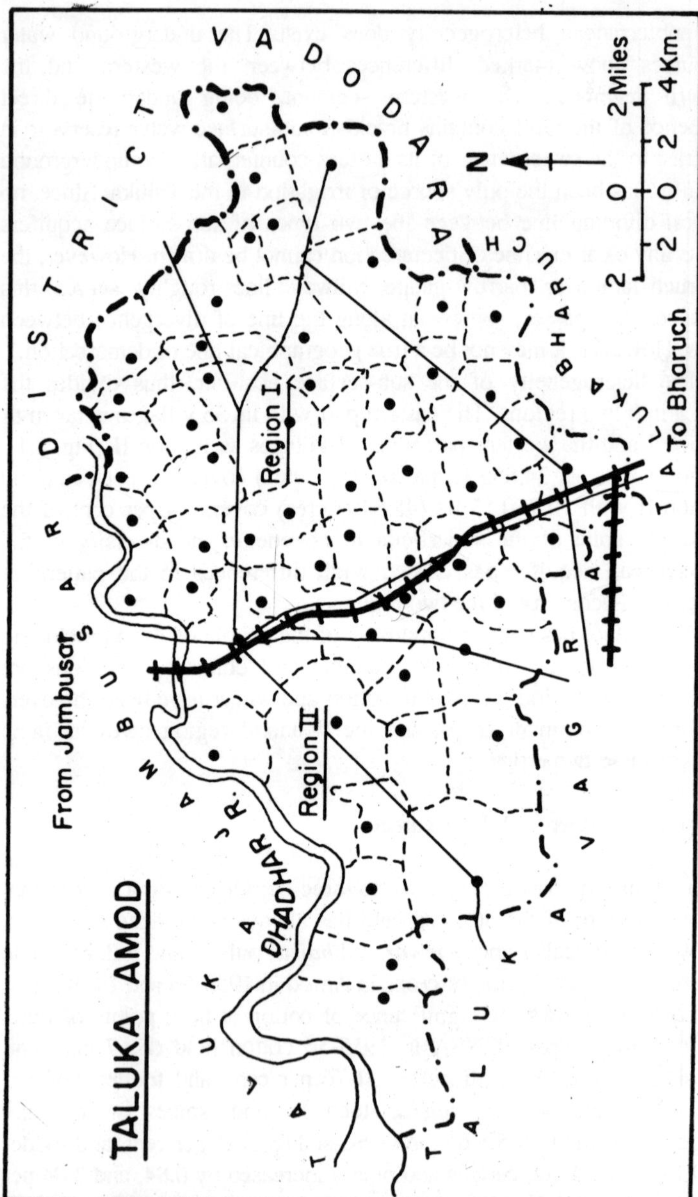

FIG 8.1: **Location and Road-Rail Links : Amod**

TABLE 8.2: Cropping Pattern of Amod — 1955-56 and 1958-59

Year	Cotton		Cereals		Pulses		Others		Fodder		Narcotics		Spices	
	Area (acres)	%	Area (acres)	%	Area (acres)	%	Area (acres)	%	Area (acres)	%	Area (acres)	%	Area (acres)	%
1	2	3	4	5	6	7	8	9	10	11	12	13	14	15
1955-56	61601	65.67	13903	14.81	571	0.61	10317	11.15	7104	7.67	68	0.07	24	.02
1958-59	58261	62.06	14694	15.65	NA	—	13320	14.19	7040	7.40	66	0.07	NA	—

the other crops in its association were mostly for the local and personal needs or under the politico-economic circumstances such as the Grow More Food Campaign.

Hybrid Cotton

Attempts were on to improve the quality and yield of cotton and other crops. Since 1955, cotton underwent successive experiments in this direction. Vijay strain, the most popular of that time, was improved upon to Digvijay, thereby its quality and yield were improved. Like Vijay, this improved strain was also adapted to the natural conditions of Amod, and its adjoining areas. Further strain improvement brought about Hybrid-4 (Shankar Char) around mid-sixtees. It served Two purposes – improved the quality and raised the yield. Its botanical parents were Gossipium Herbaccum (Gujarat-67) and American Necterites. It was much superior to Digvijay in respect of the pod, the lint value, seed and staple length. Another similar attempt brought about Varalaxmi strains. Late entrants to the cropping environment were Hybrid 5 and 6. Table 8.3 gives a comparison between the quality and yield of a few strains of cotton cultivated at Amod.

Table 8.3 shows that Hybrid-4 and Varalaxmi bear more or less the same qualities, but the former gives higher yield than the latter. Both of them are superior to the former varieties, but inferior to them in the percentage of cotton in the pad. However, the yield neutralises this defect. It is the quality of the higher per ha yield of these hybrid varieties that attracted the cultivators. Over and above, this was the advice of the Government not to grow the old varieties. Thus, in a short span of time the new strains became very popular in the taluka.

TABLE 8.3: **Comparative Quality and Yield of Different Varieties of Cotton**

Varieties	Weight of lint gr.	Staple length mm	Spinning value	Weight of Seed	Yield per ha (in kg)	% of cotton in the pod
Vijay	2.00	22.6	39	N.A.	538	39.3
Digvijay	2.50	23.1	42	N.A.	663	39.0
Hybrid-4	4.00	29.0	60	3.00 gr	3500-4000	34.4
Varalaxmi	4.00	30.5	60	3.00 gr	3000-3500	29.8

The basic difference between the old and new varieties was that the former were adapted to the prevailing natural conditions of the taluka, as they were resistant to climatic variations, pests and diseases.[2] Irrigation and chemical fertilisers were seldom used, whereas new varieties could not do without these.

Changing Cropping Pattern

The increasing cultivation of hybrid varieties of cotton displaced the old variety from most of the villages of the Taluka. Table 8.4 shows the changing cropping pattern of the taluka since 1975-76 (Fig. 8.2).

Table 8.4 shows paradoxically that the introduction of the Hybrid varieties of cotton have had the effect of reducing the percentage of G.C.A. under them as compared to 1955-56 and 1958-59. The share of land under cotton has gradually decreased from 48.94 per cent in 1975-76 to 36.22 per cent in 1978-79. However, the G.C.A. has also decreased by 2239.17 ha in 1976-77, and 2373.94 ha in 1977-78, but has increased by almost double (4705.78 ha) to that of 1977-78 in 1978-79, while the cotton area has declined to an all time low in the same year. The advantage of the decline of cotton is, however, enjoyed by cereals, *tur*, and other crops. However, *tur*, which was quite insignificant (0.61%) in 1955-56, got a cognizable place among the crops by occupying 18.25 per cent of the G.C.A. in 1978-79. This may seem anomalous. With high-yielding varieties more land could have been devoted to cotton for high profit. But this has not been the case. One reason for the decreasing proportion of land under cotton is the availability of higher quantity of cotton from a smaller hectarage, thus allowing more land to release for pulses and other highly profitable crops, for which there has been consistent governmental effort, goading the farmers to grow more pulses to meet the national deficit. But the overall decrease conceals the differential changes in the two regions. A greater fall is seen in Region II than in Region I. This can be seen from Tables 8.5 (a) and (b).

These tables show that, inspite of the similar genere of crops, the preferences of both the regions are different. Cotton remained the most preferred crop of both Regions I and II. Similarly, cereals and pulses also occupied their due shares. Vegetables did not receive much favour in Region I, but they got relatively more favour in Region II, mainly because of the nearness to the two towns, Amod and Jambusar. While Region I gave greater weightage to cotton during first two points of time, Region II gave greater weightage to cotton followed by cereals.

TABLE 8.4: Percentage Area Under Different Crops 1975-76 to 1978-79

Years	Cereals	Pulses		Cotton H-4	Fodder	Other crops	Grass cropped Area (ha.)	Total
		Tur	Others					
1975-76	24.21	4.81	11.36	48.94	3.45	7.23	37477.54	100
1976-77	26.33	6.25	8.72	48.68	3.10	6.92	35238.37	100
1977-78	24.10	7.54	9.26	47.34	4.03	7.73	32864.43	100
1978-79	26.83	18.25	7.47	36.22	3.23	8.00	37570.21	100

Source: Taluka Form XX, T.D.O., Amod.

TABLE 8.5 (a): Percentage Area under Different Crops 1975-76 and 1979-80 (Region I)

Year	Cereals	Pulses		Vegetables	Cotton	Grass	Other crops	Total
		Tur	Others					
1975-76	19.72	4.90	12.17	4.54	54.17	4.00	0.50	100
1976-77	22.94	4.06	10.56	4.24	52.70	3.63	1.87	100
1977-78	22.00	8.70	10.60	3.14	50.60	4.32	0.64	100
1978-79	22.10	18.66	8.00	4.01	42.50	3.77	0.96	100
1979-80	14.93	21.55	5.37	3.65	48.89	4.64	0.97	100

TABLE 8.5 (b):Percentage Area under Different Crops 1975-76 and 1979-80 (Region II)

Year	Cereals	Pulses		Vegetables	Cotton	Grass	Other crops	Total
		Tur	Others					
1975-76	30.20	3.91	9.64	9.06	43.22	2.70	1.27	100
1976-77	36.00	2.93	7.40	10.70	38.61	3.00	1.36	100
1977-78	28.11	5.46	6.94	12.46	41.71	3.53	1.79	100
1978-79	33.98	17.62	6.60	11.60	26.02	2.52	1.66	100
1979-80	30.38	25.47	8.21	11.23	20.14	3.82	0.75	100

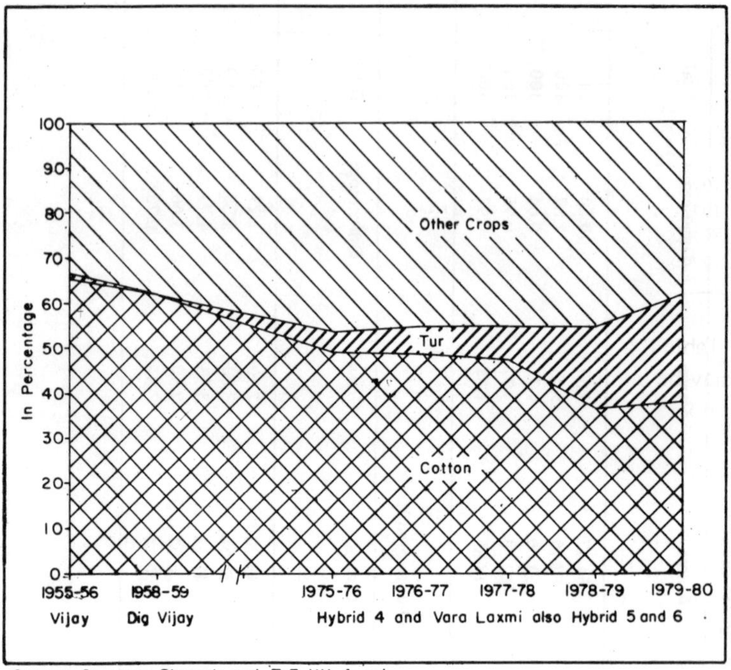

Source : Gazeteer Bharuch and T.F. XX Amod.

FIG 8.2: Taluka Amod : Changing Cropping Pattern, 1955-56 to 1979-80

Where cotton ascended gradually in Region I, Region II showed an ascent by 3.1 per cent, and a drop in cereals by 7.89 per cent. While cotton went up by 6.39 per cent in Region I in 1979-80, over 1978-79, it went down to an all time low of 20.14 per cent in Region II. A new competitor of cotton and cereals emerged in the form of *tur* (*Cajanus Indicus*) which registered an increasing trend in both regions since 1977-78 and acquired a high place in both regions in 1979-80 by occupying 21.55 per cent of G.C.A. in Region I, and 25.47 per cent in Region II. Thus cotton lost ground in the triangular competition with cereals and *tur* in Region II but maintained its relative supremacy, though a bit subdued in Region I.

Apart from the regional variations, and changes in the cropping pattern, the pattern of the whole Taluka has also drastically been affected by the change in the varieties of cotton. Table 8.6 shows the rise of other crops, which were only occupying 34.38 per cent in 1955-56, and 37.94 in 1958-59.

TABLE 8.6: Rise and Fall of Cotton and Other Crops 1955-56 to 1979-80

(in per cent)

Year	Cotton	Other crops
1955-56	65.62	34.38
1958-59	62.06	37.94
1975-76	48.94	51.06
1976-77	48.68	51.32
1977-78	47.34	52.66
1978-79	36.22	63.78
1979-80	37.89	62.11

Table 8.6 provides the evidence as to how the cotton has declined. It may be admitted on the strength of the figures in the table that the attempt to hybridise the cotton paved the way for its decline in Amod. In 1955-56 when the Vijay strain was popular, it occupied about 66 per cent of the G.C.A., its hybrid product Digvijay in 1958-59 went down by 3.56 per cent. In mid-seventees hybrid-4 and others further declined by 13.12 per cent compared to 1958-59. It is found that from 1956-57 to 1978-80 a perpetual decreasing trend set in cotton, and increasing trend in cereals and other crops. However, a slight increase by 1.67 per cent was seen in cotton in 1979-80. The sole responsibility of dethroning cotton from its leading position in Amod goes to no other crop than to the Hybrid 4 and its associates.

Factors of Change

The physical, biotic and economic factors may be attributed to be the strong reasons for the change in the cropping pattern of Amod over the years.

Physical

The region of black soil *(Kanam)* bears more or less the same characteristics as of the well-known black cotton soil. Its sticky, moisture retentive nature accompanied with the marine influences on the local climatic conditions made it an outstanding cotton cultivating area of Gujarat. An optimum natural supply of water has been ideal for successful cultivation of cotton. But the irregularity and uncertainty of rains compels farmers to have a dependable source of irrigation. Region I is privileged to have these because of sweet and soft underground water. Irrigation helped to grow all crops, but more stress was laid on the cultivation of value-added crop, cotton. This region became capable of supplying the requisites of hybrid varieties of cotton, and with promising yields the economic condition of the farmers improved. Sarbhan village of this region became an example of this success story. This village alone has 69 tube-wells. Other villages also became quite well off. Region II was handicapped in this respect because of saline underground water. No irrigation could be provided to cotton as demanded by it.

But the capacity of the soils to absorb water was limited. The top soil in Region I became saturated. The saturated soils along with the chemical fertilisers reacted in a dangerous way. On the one hand it raised the yield, but on the other it produced insects and pests inimical to cotton.

Biotic

The outcome of the large-scale irrigation took a dangerous turn. A host of insects and pests sprang up. Some of the most ferocious ones are Jassids, cotton mite, cotton bugs, leaf rollers, spotted-boll worms, pink-boll worms, *lashkari* eead. As a result wilt and root rot came up as common diseases.

The hybrid varieties are highly susceptible to various pests and diseases, the virulence of which varies with changes in seasonal conditions. They attack more intensively in wet conditions than in dry conditions. The cotton of Region I first came under the direct attack of these insects and pests. To fight with these and to save the crop large-scale spraying of insecticide became a must.[3] At every fifteen day interval the insecticides have to be sprayed to save the crop. The spraying of insecticides forced the insects and pests to migrate to the nearby cotton fields. Region II was the nearest area. The economic conditions of the farmers of this region was not as sound as that of Region I, so the needed amount of insecticides was not sprayed and the result was the destruction of cotton in this region. The insecticides were an additional burden of expenditure on the farmers' limited income.

Economic

So long as the traditional varieties of cotton, (Vijay etc.) were cultivated, more than 60 per cent of the G.C.A. of the taluka was devoted to them. They seldom needed expensive infrastructure such as irrigation, chemical fertilisers, insecticides and pesticides. ***Cotton in India***, a monograph (1961, p. 315) estimated the per acre total input in 1959-60 to be Rs. 93.09 and output to be Rs. 182.00 thus giving a margin of profit to the tune of Rs. 88.91. This favoured a profitable cultivation of cotton and thus the increase in hectarage. But the input increased with the change of strains. The hybrid varieties needed irrigation, chemical fertilisers, insecticides etc. Apart from them, the cost of seeds, the labour wages, and other inputs also showed an upward trend. Thus the expected margin of profit from the HYV crops (cotton) did not materialise. Table 8.7 gives the input-output ratio of 1978-79.

Table 8.7 provides the input-output ratio of irrigated and unirrigated H-4, Varalaxmi and Digvijay varieties of cotton. When compared with 1959-60, the net profit has gone up from Rs. 88.91 to Rs. 2050 in 1978-79 but the percentage of profit and the input-output ratio has drastically been slashed from 95.51 in the former year to 48.81 in the latter year. Similarly, the ratio has also gone down from 1:1.96 to 1:1.49 and below. Thus the increase in inputs made cotton relatively unprofitable. But this ratio may be obtained under the circumstances that each acre yields the amount given in Table 8.7 Since the biotic menace is uncontrollable even on a war footing, the given amount of yield may hardly be obtained from each village. The increasing input and decreasing output of cotton, however, prompted the farmers to opt for low input crops. The cotton, which enjoyed the topmost rank among the crops in the agricultural history of Amod gave way to cereals and pulses (*tur*) in both regions, but more so in Region II. Incidentally, the decline of cotton somewhat synchronised with the rising prices of *tur*. Table 8.8 shows the prices of *tur*.

The attractive prices of *tur*, thus allured the farmers, and the monotonous agricultural landscape persisting for long, experienced a perceptible change.

Findings

(i) Hybrid-4 and other such strains of cotton do not suit the agro-climatic environment of Amod,

TABLE 8.7: Input-Output Ratio of Different Varieties of Cotton of Amod 1978-79

Varieties	Total input per acre in Rs.	Total output per acre in Qntls.	Total output in Rs.	% of profit to input in 1978-79	% of profit to input in 1959-60	Input-Output ratio in 1978-79	Input-Output ratio in 1959-60
H-4 & Varalaxmi Irrigated	4200.00	10.00	6250.00	48.81	—	1:1.49	—
H-4 & Varalaxmi Unirrigated	N.A.	N.A.	N.A.	—	—	—	—
Digvijay Irrigated	1700.00	4.00	2400.00	41.18	95.51	1:1.41	1:1.96
Digvijay Unirrigated	950.00	2.00	1200.00	26.32	—	1:1.26	—

Source: Fieldwork and "Cotton in India" 1961, p. 315.

TABLE 8.8: Market Prices of Tur at Bharuch

Year	Prices in Rs. per Qntls.
1975-76	113
1976-77	126
1977-78	326
1978-79	326
1979-80	281

Source: Deptt. of Ag. Statistics. (Ahmedabad).

(ii) Their demands of chemical fertilisers, irrigation and insecticides are high which the average farmer is unable to afford,

(iii) Being more susceptible to pests and diseases, ruinous results have been experienced by the farmers leading (often) to near total extinction of cotton from many villages in Region II and also in Region I,

(iv) Increasing input and decreasing output prompted the farmers to opt for low input crops such as cereals and *tur*, and

(v) Successive crop failures increased rural indebtedness.

Suggestions

(1) Irrigation facilities must be provided to all villages. But the farmers should be educated to use water economically and carefully so as not to spoil the soil by waterlogging or allow the germination of pests and diseases.

(2) Financial assistance should be extended by the government on long term through, low-interest loans, to revive the lost domain of cotton.

(3) Pests and disease control measures should be taken up by the Taluka and Gram Panchayats.

(4) Adulteration of insecticides and pesticides should be checked and adulterators brought to book.

(5) The Cotton Research Institutes should find out pest and drought resistant high yielding varieties of cotton which may successfully be grown by dry farming methods.

These measures, if implemented, may bring about better all-round development of the rural areas concerned.

NOTES

1. In the light of the experiences of the farmers of Amod it is learnt that to the west of the railway line the subterrain water is brackish and to its east it is sweet and soft. Some farmers of the western part purchase water for irrigation from the farmers to the immediate east of the railway line.
2. Farmers reported that the insects and pests were unknown to the area prior to the introduction of hybrid varieties, chemical fertilisers and irrigation.
3. Village Sarbhan used helicopters to spray insecticides.

9

LAND-USE AND RURAL DEVELOPMENT
A Case Study of Navi Jithardi Village

C.M. GUPTA

THE STUDY of the relation between land-use and rural development has recently become an integral part of geographical study. India is a representative of villages. Therefore its social, cultural and economic development depends on the proper utilisation of its land resources spread over in the villages. The village lands help us in the production of food, fat, fibre and fodder required for our human and animal population. The proper use of land is the fundamental factor in an area's development. Merely a huge area of any village is not helpful for its economic development, but what is more important is the proper use of land and proper land-use planning. These two aspects are key to the prosperity of the villages.

The present paper is an attempt to make a micro-level study of the relation between land use patterns and rural development in Navi Jithardi village in Karjan Taluka of Vadodara District in Gujarat.

Geographical Background

Navi Jithardi village (22° 2′ N. Lat. : 73° 11′ E Long.) has a total area of 683.8 acres, with a population of 703. The village is connected by roads with its neighbouring villages like Kandari in the North, Dhawat in the East, Bharthali in the South and Juni (old) Jithardi in the West (Fig. 9.1). It is about 4 km South-East of Miyagam-Karjan Railway Junction on the Baroda-Bombay Western Railway line.

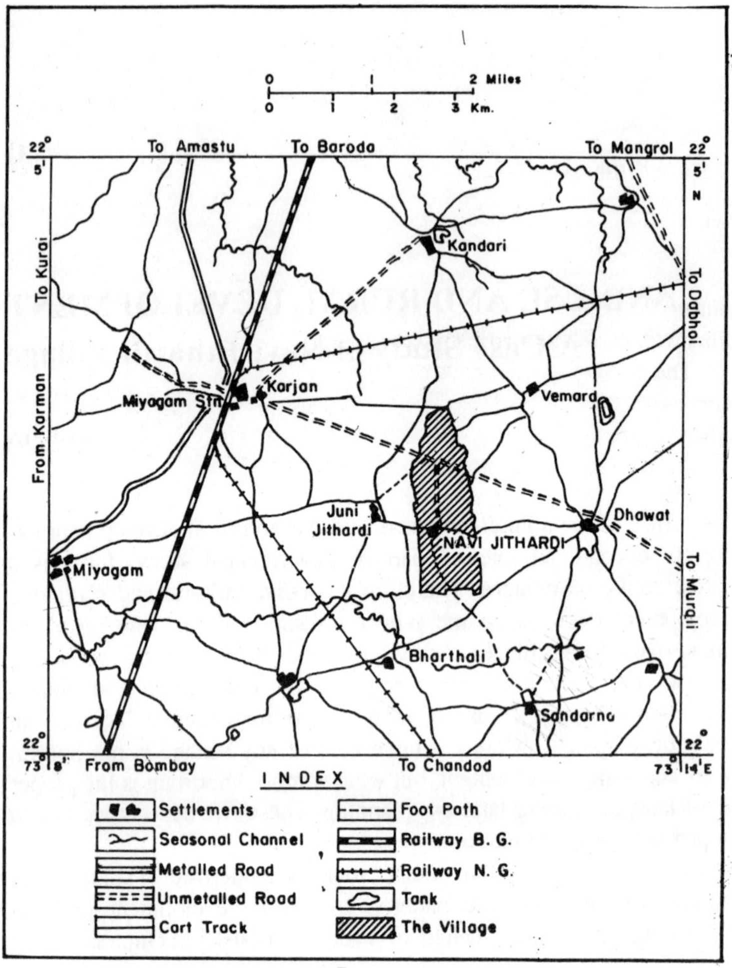

FIG 9.1: Navi Jithardi : Regional Setting

The entire village presents almost a flat level plain remarkably homogeneous in character with the exception that at certain places, the monsoon *nullas* have formed some depressions in the south-western portion of the village.

The village comes under BShw category of Koppen's climatic classification; as such nearly 98 per cent of the total annual rainfall is received during the period from June to October from the South-west monsoon, as it is clear from the Table 9.1.

TABLE 9.1: Average Rainfall

Station	Annual Rainfall	June-Oct.		Nov.-Feb.		March-May	
		mm	%	mm	%	mm	%
Miyagam Karjan	1107.8 mm	1086.1	98.04	16.4	1.48	5.3	0.48

Source: The Executive Engineer, Irrigation Division, Baroda.

The mean maximum temperature of the village goes up to 32°C during the summer season, while in the winter season it comes down to the mean minimum of 19.7°C.

The major portion of the village has the best deep black, clayey and somewhat calcareous soil ("A"), locally known as *Kali* (Black). It is ideally suited to cotton cultivation. It has a tendency to become very adhesive when wet and to develop deep cracks when dry (Fig. 9.2).

Owing to its high clay content it becomes unworkable during heavy rains. Its moisture retaining capacity is so high that it can produce a winter crop without irrigation after a normal rainy season. Therefore, in the absence of irrigation facilities, agriculture is entirely dependent on rainfall, and a failure of the monsoon rains involves the loss of both the *Kharif* and *Rabi* crops.

The medium quality ("B") land is medium black and covers a small area in three small scattered patches. This is the *goradu* soil. This soil is also not irrigated. Portions of this land are used for cultivation of groundnut and for grazing purposes.

The poor quality land ("C" – the *Besar*) is used as a grazing land and the cemetery of the village.

General Land-use

These Physical conditions influence land-use a great deal.

Table 9.2 gives a summary of the proportions of the village land devoted to various uses.

High soil fertility and a flat surface have resulted in a very high percentage of cultivated land higher than State and national percentages. No land is kept fallow or given a period of rest. The land is cultivated continuously year after year leaving some land for houses, roads, tanks, etc., and some portion is utilised as the village grazing land. The proportion of wasteland etc., is higher here as compared to the Taluka, but the area under grazing and village common is much less.

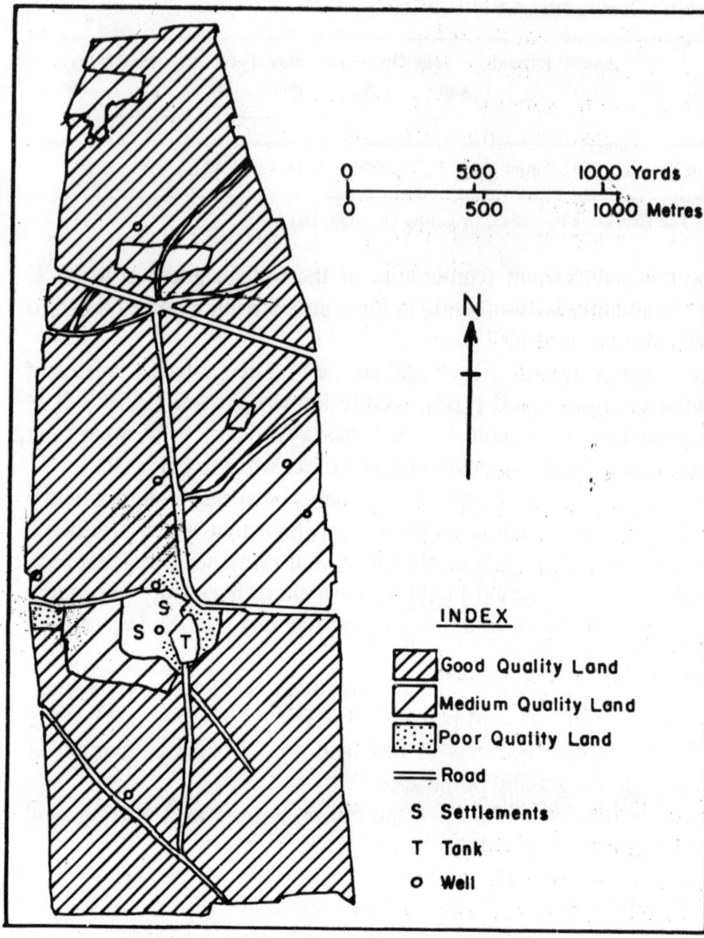

FIG 9.2: Navi Jithardi : Land Classification

In either case it can safely be concluded that some more land can be brought under cultivation or other profitable uses provided the constraints are removed.

Among the crops, cotton takes the lion's share. This is the dominant cash crop of the village. It occupies 89.40 per cent of the net sown area. The rabi crops of *jowar*, wheat and castor are insignificant.

Wheat is grown singly, while castor is usually cultivated as a mixed crop in combination with cotton, *tur* and *jowar* (Fig. 9.3).

TABLE 9.2: General Land-use

Use of land	Area in acres.	Percentage to total areas of:		
		Village	Taluka	State
Net Sown Area	614.1	89.47	86.40	53.3
Fallow land	—	—	0.91	5.7
Forest	—	—	—	8.4
Land not available for Cultivation (Waste-land, Settlements, Village Road & Cart Track)	30.9	4.97	3.20	23.7
Other Uncultivated land (Grazing land, Cemetary and village common)	38.8	5.56	9.50	8.9

Cropping Pattern

Thus the cropping pattern of the village is quite simple. The cotton fields are usually much larger than the rice and *jowar* fields. Mostly the patches of *jowar* fields are seen in the southern and eastern parts of the village and some are found towards the central and in the northern portions. Wheat and castor fields are scattered all over the village. Close to the village site, paddy is grown in small fields where the farmers can pay greater attention.

TABLE 9.3: Cropping Pattern

Crops	Area in Acres		Percentage of net sown area	
	Village	Taluka	Village	Taluka
Kharif Crops:				
Cotton	549.1	86579.0	89.40	67.62
Fodder (Batu, Grass, and Sudhia)	24.9	N.A.	4.05	N.A.
Rice (Paddy)	18.6	4062.6	2.93	3.17
Tur (arhar)	11.9	2920.8	1.99	2.28
Groundnut	9.6	1063.6	1.63	0.83
Rabi Crops				
Jowar	38.2	16339.8	6.19	12.76
Wheat	5.1	2269.0	0.81	1.77
Castor	2.2	N.A.	0.32	N.A.

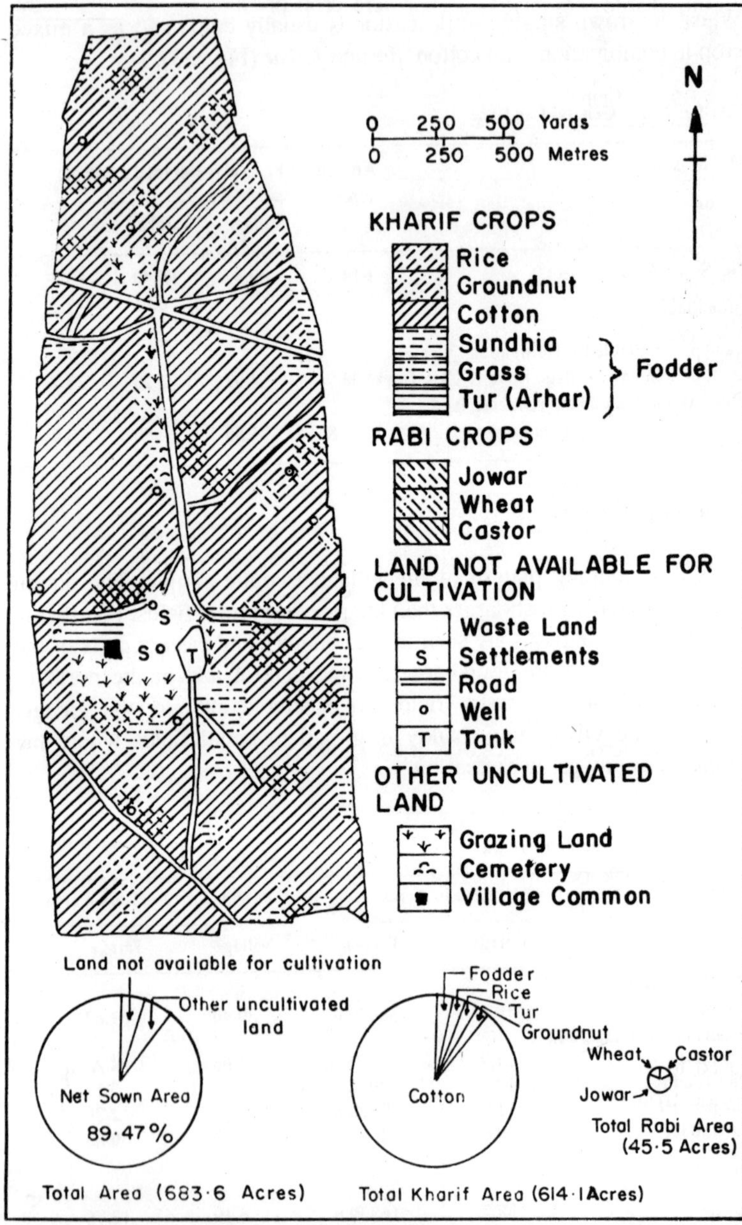

FIG 9.3: Navi Jithardi : Land Utilisation

Crop Productivity

Crop production of this village is high due to fertile soil and adequate rainfall. Table 9.4 shows the per acre crop productivity for 1965 and 1980.

TABLE 9.4: Crop Productivity in kg/Acre

Crops	1965	1980
Cotton	100—150	175—200
Fodder	200—250	250—300
Rice (Paddy)	150—200	200—250
Tur (arhar)	200—250	300—400
Groundnut	200—300	400—500
Jowar (Rabi)	150—200	200—275
Wheat	350—400	450—550
Castor	100—150	150—200

Source: Personal Survey.

The above figures make it clear that the productivity is not only high but has increased to great extent since 1965 due to the impact of the Green Revolution. Adoption of agricultural innovations like the use of high yielding varieties of seeds, chemical fertilisers (apart from the farm manures) and pesticides. Also better prices for the high-yielding variety of cotton crop has added to the impetus. The "Shankar-4" variety of cotton is a very famous and profitable cash crop and this crop alone has changed the destiny of Navi Jithardi by making it as one of the prosperous villages of Karjan Taluka.

Livestock

The livestock occupy an important position in the rural economy of this village. They supply most of the motive power required for agriculture. The cows, and the buffaloes are important as producers of milk which is of vital need in the diet of the people of Navi Jithardi, who are mostly vegetarians. Milk is also needed for making ghee, butter, mava, curd and various other preparations which are commonly used in the diet by the people as well as sold for cash. The dung and urine from livestock are mostly used as manure. They provide hides and skins, bones, hooves and horns which are of considerable importance to the economy of this village.

Average yield of milk from cows is reported to be 1400 litres, and from buffaloes 1800 litres. The total annual milk production was 1,314,00 litres in 1979-80. This milk brings a good earning to the people of the area. This has become a permanent feature in the economy of the village. Therefore, for the propagation of standard breeds, the Govt. have opened key village centre at Karjan for efficient breeding. The Govt. have also opened one Artificial Insemination Centre at Karjan, and the response from the rural people is very encouraging. The Patel farmers of this village are always ready for carrying out agricultural experiments in their fields.

Grazing Facilities

Navi-Jithardi village has a common grazing ground lying to the north-east and south-west of the village tank near the settlement. During rainy season the grazing ground remains green and animals remain satisfied with grass but in hot dry summer, because of water scarcity, the grasses wither and dry up and the villagers find it difficult to maintain proper feeding of animals.

The farmers do not devote large areas to fodder crops. For this they depend mainly on the by-products of crops already grown and meet their additional requirements by purchasing fodder from the neighbouring villages. Jowar is a highly productive crop and is fit for silage also. For Navi-Jithardi village, it is a rabi crop. The green and dried stems of this crop are locally termed as 'Batu' and 'Pura' respectively.

Apart from Jowar, the straws of wheat and rice are also used as dry roughages for the village cattle population.

Economic Activity

Nearly 92 per cent of the people of Navi-Jithardi are engaged in ploughing, sowing and harvesting. Amongst the rest, six persons prepare *Mava* and send it to Bombay through fast trains from Miyagam Karjan Jn. There are four general shop-keepers, three tailors, two barbers, three carpenters and ten persons go to the neighbouring Miyagam Karjan village for working in a ginning factory.

Eighty to 85 persons are engaged in animal husbandry. Almost 360 litres of milk is produced and it is daily sent to Baroda Dairy through the dairy contractor. The supply of milk runs short in summer because the milk capacity of cows goes down due to little grass to feed on. Milk producing and selling brings daily income to the farmers and this activity has recently developed with the help of co-operative banks.

The people sell their cotton at good price in Karjan, which is one of the larger village of Vadodara District. Karjan is semi-urban and also the headquarters of the Taluka.

Because of the agricultural prosperity the village has facilities like phone. Almost 30 to 40 per cent of the houses keep cycles for going from village to Karjan. Eighty per cent of the houses keep radios for listening to the agricultural programmes broadcast every day in the evening. Some of the Patels of this village, whose children are in foreign countries, even possess TV sets.

The *Gram Sevaks* and the *Gram Panchayat* members of the village look after the sanitation of the surroundings. They take keen interest in the social and cultural activities of the village. The Post Office and hospital facilities are fully shared by the villagers.

There is one primary school and one middle school in which 170 boys and girls study. The village is electrified and water is supplied through taps. The village has one tube-well and two flour mills.

Conclusion

Looking at the general condition of the people of the village it can very well be concluded that they are quite well-off, much more advanced and developed than the rice growing farmers of the Ganga Plain. They can be compared with the developed farmers of Punjab and Haryana. The main reason for this is the practice of growing cotton as a cash crop. The growing of cotton and its disposal for profit have transformed the subsistence economy of the farmers into a commercial economy. This has changed their whole attitude to life. With the surplus earned they can have the modern amenities of life, take full care of their health, send their children for study in foreign lands, and even spend on luxuries and on religious ceremonies. Besides, they can invest in improving the productivity of their farms and invest in other productive pursuits like trade, commerce and industry for which the people here are famous.

10

USE OF REMOTE SENSING DATA FOR RURAL LAND-USE SURVEY

S.B. DESHMUKH

IN HISTORY of geography it has been presumed that maps are indispensable tool of geography. They communicate interrelationship between man and environment. Their prime purpose is to acquaint us with physical as well as cultural features on the surface of the earth. This information is collected by visiting the places and surveying them. Obviously completion of a map takes a long time. This limits the reproduction of maps at shorter intervals. The process itself is a prolonged one.

But time will not wait for us. The landscape is changing very fast. Yesterday's invention becomes too old today. We must cope with the changes. We, the geographers, are always in search of data because we cannot take a single step without it. We try to get information basically from maps. We feel that such collected information is very old for today's need, but because of lack of alternatives, we use the same information which will not give up to date results and will cause errors in the planning strategy for a region. This shows incompetence of the map in providing up to date information for our planning strategy which raises doubts in our mind about the usefulness of old maps in planning.

To cope with the needs of researcher and also the planner, there is a necessity to find out an alternative so that up to date information could be available. Geographers can now look forward to aerial photographs and satellite imagery.

Aerial photographs and satellite images cope with the present needs of researcher and planner. Presently, we are trying to make use of aerial photographs for our research work and planning exercises. But in India availability of aerial photographs at the desired time is a great obstacle. Hence the most recent source available at the moment for our work is satellite imagery. We can make use of it to assess the regional resources and disparities by preparing up to date land-use maps. These maps will serve as competent base maps for comparative study of different regions which will be helpful in bringing out regional disparities at meso as well as micro levels. The imagery will not only solve the difficulty of up to date data bank but also provide information and details of smaller areas which is necessary for meso and micro-level planning.

Remote Sensing

Remote sensing is the science and art of acquiring information about material objects from measurements made at a distance without physical contact with them. This technique of remote sensing is one of the tools for gathering data. The main advantage is that it covers very large areas (185 × 185 km^2 in one format) of the earth's surface from a perspective and format which facilitates the study of the objects and their relationships. In remote sensing, information may be transmitted to the observer either through force fields or electro-magnetic fields, and in particular through the spectral, spatial and temporal variations of these fields. Therefore, in order to derive information from these field variations, one must be able to measure these variations and relate these to those of known objects or materials. For example, one desires a map showing all of the water bodies of a certain region of the earth; it is clear that one cannot sense the water directly from space craft's altitudes, but only the manifestations of the water bodies which exist at that height. These manifestations in the form of electro-magnetic radiation must therefore be measured and the measurements analysed to determine which points on the earth contain water and which do not. In this way we can utilise the imagery for preparation of rural land-use maps. This imagery or the land-use map prepared by this imagery will give the spatial distribution of land-use at a particular time. This method of preparing rural land-use maps is more or less similar to conventional aerial photography. This has now got wide importance among both the developed and developing countries of the world.

Preparation of Land-use Maps

The methodology of preparation of land-use maps is divided into two stages:

1. Pre-operational :	Involving the careful selection of the appropriate interpretation techniques and imagery.
2. Operational :	In which the actual land-use map is produced.

All aspects of both stages are considered to be relevant. To achieve optimum results in the final map, adequate attention must be paid to the planning, selection and preparation of imagery and techniques before the interpretation process commences.

Pre-Operational Stage

Planning: This is the initial phase of methodology to establish the basis for the whole mapping operation, and entails the detailed clarification of the purpose of map and a statement of objectives which must be fulfilled in order to produce the desired result. These in turn influence decisions about the most appropriate scale and detailed specifications of the final map. Both require careful consideration of possible imagery, base maps and other reference materials. Once the objectives have been defined, standard pre-processed LANDSAT/MSS imagery can then be selected for the use. The imagery should be selected so that the seasonal variations in vegetation cover can be utilised in accurate delineation and identification of land-use boundaries and categories.

The recommended standard pre-processed imagery for rural land-use mapping consists of:

One	1: 1,000,000	false colour composite transparency.
or One	1: 1,000,000	Black and white positive transparencies of bands 4, 5 and 7 for producing colour composites.

In addition

One	1: 250,000	colour composite print
or one	1: 250,000	paper print of band 5.

Preparation: This phase involves the procedures necessary for providing a satisfactory base for the detailed interpretation of land-use in the operational stage of mapping. The first step necessitates a preliminary interpretation of the standard imagery available. The accuracy of interpretation is not only affected by the quality and seasonality of the imagery but also by the interpreter's reference level, the collected material available and the visual optical enhancement techniques that can be employed. Adequate care should be taken to ensure that the interpreter spends sufficient time in becoming familiar with the nature of the MSS imagery and the spectral responses of the region as the level of interpretative skill at this stage tends to influence the framing of the land-use classification scheme, the selection of visual enhancement equipment and the type and amount of additional pre-processing. The familiarisation with the imagery should be achieved by considering relevant collateral data about the region, in addition to the methods used in acquiring and processing the LANDSAT/MSS data. Once the interpreter is accustomed to the imagery a preliminary land-use classification scheme can be evolved. This should be as broad as possible within the limitations imposed by the final mapping scale, the image scale and resolution of the spectral responses of the vegetation cover. In addition, it involves the integration of relevant collateral material, the interpreter's reference level and guidelines from other similar land-use classification systems (the U.S. Geological Survey circular 671 can give best guidance in this regard).

After considering all available techniques and imagery, those most appropriate for use in the operational stage of the map production may be selected. In addition to this, other supplementary materials and equipments that may be used to assist interpretation include air-photo mosaics at various scales, slides of portions of selected areas of LANDSAT/MSS imagery for viewing with 35 mm or 70 mm projectors, and relevant reports.

Operational Stage

The first major step in the operational stage consists of the detailed interpretation of the selected LANDSAT/MSS imagery and the production of the preliminary rural land-use map. In the detailed interpretation, colour is used as the main image interpretation element in the detection, recognition, delineation and classification of the different land-use categories. The other interpretation elements, e.g. shape, size, texture, site, etc. are all used to varying degrees depending on the type of land-use being investigated. Again, the potential accuracy level depends on the interpreters' reference level with regard to his training in image interpretation, his knowledge of specific related subjects and his familiarity with the spectral characteristics of the LANDSAT/MSS imagery. The colour composite transparency of 1: 1,000,000 with rear illumination from a light-table are initially viewed under monocolour magnification to detect areas of similar colour. After being familiar with the colour and resolution characteristics, areas of similar land-use are identified by extrapolating from the colours of known land-use areas. These can be identified from various sources including relevant reports, topographic and thematic maps of the parts of the region and air-photo interpretation of available aerial photographs or mosaics. The location of the similar areas can be facilitated by the use of a specially prepared transparent base map containing selected topographic features at a scale of 1 : 250,000.

Further assistance in the interpretation of difficult areas can be achieved by other optical enhancement techniques, especially slide projectors, and microfilm readers which may be used to optically enlarge imagery of certain portions of the regions which require further investigation. Also the spectral responses recorded on the imagery acquired during different seasons (multi-date imagery) should be viewed as an additional aid in clarifying boundaries and categories.

Once the interpretation has been completed, the boundaries and categories can be transferred to a transparent base map from 1: 250,000 colour composite print or black and white enlargement of band 5. The recommended based map scale should be the same as the print containing the integrated data (i.e. 1 : 250,000) and it should contain sufficient information to permit the location of land-use boundaries in the field.

Other aspects involved in the preparation of preliminary land-use maps, include the determination of the minimum size of the area of categories which should be mapped. The generally accepted minimum size, based on a map scale of 1 : 250,000 is approximately 25 hectares (60 acres) which represents an area of 4 sq. mm on the map. As a consequence, areas of land-use smaller than this should be merged in the surrounding category unless there are particular circumstances which necessitate their inclusion on the map. For example, certain areas of irrigated land may have special significance and may be easily identified and mapped. The appropriate symbols and colours for each category should be added to the map to enable easy identification of areas in the field. Finally, the map should be checked for omissions in the interpretation phase and in the process transferring data to the base map. This can be achieved by first over-laying the transparent map on the 1 : 250,000 print used in the interpretation and then systematically checking within each graticule of a selected grid system, e.g. 1 : 50,000 topographic map system.

After completing the work in the operational stage, the assessment of accuracy of interpretation of the land-use from LANDSAT/MSS can be done by ground verification (sampled study).

Final Map

The production of the final map involves its presentation in a complete state, ready for application in the tasks for which it is meant. It also includes the normal cartographic procedures of devising a satisfactory overall layout, the design of the map incorporating the classification system and the selection of lettering style, symbols and colours. The main aim of the colour scheme should be to make land-use maps prepared from this data source as compatible as possible.

Once completed, the final map can then be used as a final 'colour rough' for colour printing or it may be used directly as the base map for further investigation within the region. An important addition to the final map should be a report detailing the purpose, objectives and specifications of the work as well as possible explanations of the various pre-processing, interpretation, classification and ground verification procedures adopted during the production of the map.

The novelty of the technique is that the maps obtained are most objective and accurate offering reliable information for up to date assessment of the regional resources and their potentiality. Hence it is felt that the satellite imagery will prove the most useful and effective tool of all other geographical aids in getting the data, for geographic study and planning for rural development.

PART III
GEOGRAPHY AND RURAL ECONOMIC ACTIVITY

11

ROLE OF SOCIAL FORESTRY IN THE RURAL DEVELOPMENT OF GUJARAT

M.N. VAISHNAV

Philosophy of Rural Development

RURAL DEVELOPMENT has different approaches for different people. According to Dr. Schumachar, application of an appropriate technology largely based on locally available raw materials and local skills aimed at producing wage goods would be the right way for rural development. For our country mass production by few "is not as useful as" production by the masses. Rural Development in the true sense would be achieved only if the gap between the rich and the poor in rural areas is narrowed down.

First and foremost step in rural uplift is to make rural folk more aware about their environment so that a proper atmosphere for development of arts, craft and utilisation of natural resources can be created. For proper rural development productive agriculture and proper land-use is of prime importance. Optimal land-use is the prerequisite, more so, for rural areas which being farther away from industrial growth centres have to act more as suppliers of raw material as well as labour. For proper welfare of the rural populace adequate agricultural production, healthy livestock and raw material for cottage industries being essential, those aspects which can contribute to support agriculture, animal husbandry and cottage industry would be very useful for ideal rural development. Trees play a very important role in helping agriculturists, supporting animal husbandry and creating job opportunities. They have a profound influence on environment, and through trees and vegetation only we can achieve multiple land-use.

Importance of Trees and Vegetation

Help rendered by trees to agriculturists is manifold. Amongst a direct help available from trees and vegetation, fuelwood can be considered the most important. There was a time when villagers could get their fuelwood supply close at hand but it is no longer so. Fuelwood is a non-monetised commodity for the rural poor. Fuelwood gathering is the activity that has done the greatest damage to the environment and its impact has been visible even on Reserved Forests which are fast getting thinned out. Another direct help rendered by trees to the agriculturists is by way of supply of small timber for rural housing, agricultural implements and raw material for cottage industries. Role of fodder and grasses provided by trees and vegetation is also very beneficial for the agriculturists and rural folk for whom cattle are important not only for power but also for milk and dairying which provide them supplementary income and occupation. Leaves and litter available from trees and vegetation are of immense value for the farmer as they provide valuable manure that maintains the biological health of the soil and consequently its productivity. In this context, it would be worthwhile to mention about cowdung that can be gainfully utilised as manure if adequate fuelwood is made available to agriculturists and rural folk. Other produce like fruits, *bidi* leaves, fibres, flosses, gums, medicines, lac, etc., that are derivable from forests provide a very good source of supplementary income and occupation to Indian farmers. Amongst the indirect but profound benefits that the agriculturists and rural communities derive from forests are the protective role that is played by forests and vegetation in preventing soil erosion, reducing desiccation, stopping sand blows and saline winds and increasing infiltration of precipitation, etc. It is this aspect of forests that has given it a role of foster-mother of agriculture. Forests provide shelter as well as insurance to agriculture. From the point of view of job opportunities, the influx of rural population into urban areas upsetting entire planning and giving rise to a host of evils, can also be remedied by providing job opportunities in the countryside in agriculture, forestry, animal husbandry and cottage industries that need lower inputs and need low level technology.

The foregoing paragraphs would clearly indicate the need for integration of forestry with other land-uses. China is an outstanding example where forestry has been very successfully used as a support to agriculture.

Social Forestry

Social Forestry, that is, the practice of forestry through the people, by the people, for the people has an immense potential to benefit the weaker sections of the society through self help, self employment, and through meeting their minimum needs of fuelwood, fodder, small timber, and fruits. It is a mechanism that can marry land and labour without the least strain on the environment and with the least possible inputs of finance and technical know-how.

Gujarat Scene

In Gujarat, rural population accounts for 75 per cent of the total population and nearly 40 per cent of the people are below the poverty line. Since the resources are limited we must aim at schemes aiming at rapid progress over vast areas involving masses rather than the schemes benefiting few. Moreover, the gap between rural and urban incomes is often as much as 400-500 per cent and that results in an exodus to the cities, causing slum problems and allied evils. Continuing industrialisation without adequate attention to the rural scene has created serious imbalances. Due to poor productivity of many of our forest areas, coupled with other factors, like deforestation and scarcity conditions in several parts of the State, the gap between the demand and supply of forest produce has widened.

A beginning was made to involve the community in forestry some time in 1969-70 in Gujarat by inducing cultivators to grow trees. Simultaneously attempts were made to extend the horizons of forestry by taking up tree planting along roadsides and obtaining community lands like *gauchers* for tree planting. As the move received a good response, a more ambitious programme of social forestry including raising trees along roadsides, canal banks, *gauchers* etc., was taken up and the practice of farm forestry was also given a greater thrust. Looking to encouraging results and favourable support from the rural as well as urban community, an ambitious programme known as Community Forestry Project aiming at the coverage of 1,00,400 ha. of tree planting on roadsides, canal banks, along railway lines, in degraded forests, and *gauchers*, and popularising raising of farm wood lots and self-managed village woodlots was launched with the help of the World Bank in April, 1980. The project is to cost a sum of over Rs. 65 crores during its first phase covering a five-year period. It also envisages provision of 10,000 smokeless *chullahs* and 1000 improved crematoria as incentives for the community participate in the project. This move is going to result in savings of fuelwood to the tune of 15,000 tonnes annually.

The Community Forestry Project that has been launched with the financial assistance of the World Bank has put Gujarat on the World map in the field of Social Forestry because of the results achieved and the public enthusiasm generated therefrom. The project is expected to produce 3,86,000 tonnes of fuelwood and 4.8 million poles and bamboos annually from project year 10. The project is likely to generate 46.9 million mandays of labour over a period of 5 years. Apart from all the tangible gains narrated above, the project would have a major environmental impact by way of soil conservation and, in some cases, restoration of soils. It will have a tremendous educational impact too because of the ideas like school nurseries, farmer nurseries, tree organisations, involvement of womenfolk, etc., and this atmosphere would help create much better understanding about trees and environment, which in turn would be useful for development in other fields too.

Mahatma Gandhi had desired that all villages must have their own arrangements about getting their own fodder, fuel, and timber. Gujarat is sure to provide the leadership to the country in achieving his cherished hope. The activity of Social Forestry while reducing the gap between supply and demand of forest produce and ameliorating the environmental conditions will also help in the revival and stimulation of rural arts and crafts and improve the quality of life. The dedicated foresters with the support from Government and the community are bound to bring very encouraging results.

12

RURAL INDUSTRIALISATION IN A DROUGHT-PRONE AREA

M.G. BHASIN

ADMITTEDLY OUR rural areas have continued to remain at "bullock-cart" stage in spite of the planned programmes initiated ever since 1951. Judged from all points of view these are backward problem-areas. Their development problems are massive in size, complex and complicated in nature. Hence rural poverty has not shown any signs of abatement. In reality the problem herein is not only unemployment but also under-employment which means work for only six months in a year!

Included in these rural areas of India are the drought-prone areas where the development problems are all the more severe, alarming and challenging. These areas are identified and delineated in various ways by various authors. One may gainfully say that these occur at situations where water losses due to evapo-transpiration exceed the rainfall. Their genesis may be easy to visualise but is certainly difficult to explain!

On the basis of intensity and past occurrence of droughts, the Government of India in 1970-71 selected six project districts in four States under DPAP, viz. Jodhpur and Nagaur in Rajasthan, Solapur and Ahmednagar in Maharashtra, Bijapur in Karnataka and Anantpur in Andhra Pradesh.[1]

The Study Area

The present paper concerns with Solapur District (17° 10′ to 18° 32′ N.

and 74° 42′ to 76° 15′ E; 15021 sq. km). This district has a long history of drought-caused famines, but their intensity and frequency varies from taluka to taluka (see Table 12.1)

TABLE 12.1: Frequency Variation of Scarcity Conditions in Solapur District

Grade	Estimated frequency of scarcity condition	Talukas
A	Once in 3 years	Karmala, Sangola
B	Once in 6 years	Madha, Pandharpur, Malshiras (partial)
C	Once in 10 years	Mahol, Mangalwedha, Akalkot, North Solapur, South Solapur and Barsi (partial)

This condition is associated with its tropical situation and erratic nature of monsoon circulation. The following local characteristics have been noted, from the available records:

(i) normal onset of monsoons is followed by 2-3 intermittent dry spells,
(ii) early start of monsoons is followed by continuous dry spells,
(iii) delayed onset of monsoons,
(iv) early withdrawal of monsoons, and
(v) prolongation of monsoons up to middle of November.

Solapur District lies central to the two project Districts of Bijapur (Karnataka) and Ahmednagar (Maharashtra). It has been formed by merging areas from Pune, Ahmednagar and Satara Districts, at first in 1839 and later in 1869. Presumably, its name is derived from two words *Sola* meaning sixteen and *pur* meaning village. According to the Census there are 948 villages in its 11 Talukas. Although Solapur is pre-eminent for the manufacture of Jacquard *chaddars*, wall-hangings, tapestry among textiles, and for *bidi* making, yet its economic well-being is disadvantaged by industrial inertia. It has affected both industrial production as well as labour productivity. For convenience its industrial activity may be characterised as under:

(i) The industrial activity is concentrated at Solapur, the fourth largest city in Maharashtra State,
(ii) The fortune of the existing industries is dependent on the raw materials, whose supply is in the hands of an unstable market, and

(iii) The labour is traditional and mostly illiterate. Their physiological capacity and mental attitude are very much affected by climate. Absenteeism and labour costs are the two major bottlenecks for the entrepreneurs.

In levels of economic development the district is below the all-India average but in Maharashtra State it ranks eighth, the first seven being Greater Bombay, Pune, Thane, Jalgaon, Ahmednagar and Nagpur.[2] So, though it is a backward district, it is not too backward. To accelerate economic development attention is being given to its industrial development. The Solapur District Industries Centre is operating an "Action Plan" and the Bank of India, the lead bank for the District, formulated Solapur District Credit Plan for 1980, 1981 and 1982.[3]

Limited Aims

There are really three sets of basic needs for rural development – those of the rural area and their inhabitants, those of the nation as a whole, and those of the international system. To reach the rural areas and their inhabitants importance has been given to rural industrialisation. This is one important element, one sector, or a process in the rural development. Its two major aims are:

(i) Increasing industrial production, and
(ii) Creating jobs in rural areas.

Further aims can also be distinguished as they are oft-spoken or are implied in the nature of plans for industrialisation:

Major Aims

(i) Improving rural economic structure,
(ii) Industrial decentralisation,
(iii) Improving inter-regional income distribution,
(iv) Supplying regional markets, and
(v) Stopping migration to cities.

Minor Aims

(i) Using local resources,

(ii) Increasing industrial production,
(iii) Development of Entreprenership,
(iv) Creating jobs, and
(v) Improving quality of labour.

Once the aims and objects are clearly defined and stated it is relevant and logical to seek enquiry into the following questions:

(1) What industries and crafts are located in the selected area?
(2) What factors determined the choice of their location?
(3) What industries and crafts are selected for assistance and promotion?

Ideally an industrial geographer is best suited to describe and explain spatial patterns of manufacturing and had competence enough to suggest potential points and areas of manufacturing, industries and crafts for assistance. But sadly enough he goes without recognition.

Imbalances in Strategy

All types of industries – large, medium, small-scale and cottage industries – can be an important part of the economy of a rural area. But small-scale and cottage industries may be more suitable.

Taking only the small-scale industries in the district, a steady rise in number was registered from 260 in 1965 to 867 and 1385 in 1971 and 1975 respectively on to 3163 in 1979. Furthermore, their concentration varies from type to type; highest being in textiles and lowest in plastics (Table 12.2).

Besides, there are six sugar factories in the district. These are located at Akluj, Malinagar, Sadashivnagar, Pandharpur, Bhogavati and Shreepur. Three spinning mills have been sanctioned for the district of which one has been commissioned at Valsang. The handloom industry is scattered at places such as Karajgi, Tolanur, Maindergi, Vagdari, Valsang, Sangola, etc.

Figure 12.1 shows the location of growth centres A and B as well as the Mini Estates being established. Secondly, regional patterns have been brought out by rating Talukas by simple ranking method; and assigning values as 5, 3 and 2 for Growth Centres A, B and Mini Estates respectively. The emerging pattern points out to a *differential approach* rather than a *unifying approach*. This situation seems to be the outcome of following factors:

TABLE 12.2: Composition of Industries

Sl. No.	Type	No. of units	Ranks	Employ-ment	Ranks
1.	Textiles	2624	1	17,500	I
2.	Ready-made Garments	100	3	166	V
3.	Wooden Shuttles	6	9	52	X
4.	Wooden bobbins	4	11	50	XI
5.	Chemicals	28	6	166	V
6.	Engineering	172	2	1735	II
7.	Food Processing	80	4	495	III
8.	Printing	74	5	396	IV
9.	Forest-based	21	7	110	VI
10.	Cement-based	6	9	63	VII
11.	Rubber Goods	5	10	55	IX
12.	Leather	4	11	63	VIII
13.	Servicing	10	8	49	XII
14.	Plastics	3	12	21	XIII
15.	Miscellaneous	26	—	79	
		3163		22,084	

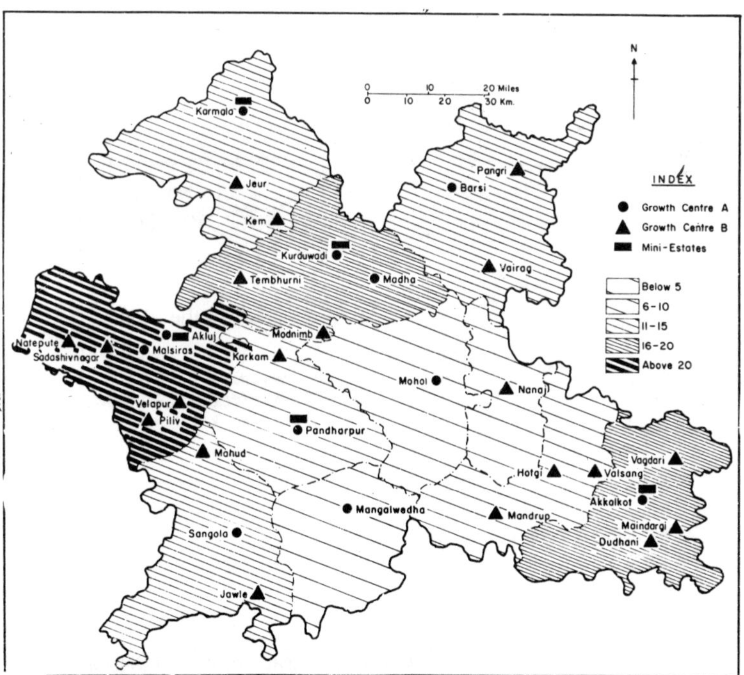

FIG 12.1: Solapur District : Industrial Pattern

(i) The plans are not being prepared from grass roots. In fact the rural industries are being planned within a national development model. As such the plan is less real, less effective and at times rigid. It has increased regional disparities,

(ii) The industrial affairs are being increasingly subordinated to politico-administrative direction and controls. To name a few, we have industrial licensing, company law, foreign exchange regulations, price controls, restrictions on the supply and distribution of essential commodities and laws about monopolies, etc.

(iii) "Educated incapacity" of the planners and administrators who do not know how to plan comprehensively, completely and effectively. For example, to plan effectively the development in rural areas should have foundations in "self-reliance" or "self-help". But the drought-prone area of Solapur District is caught in a vicious circle (Fig. 12.2). Hence, there is a real fear that this approach may not lead us to meaningful employment for all-round development.

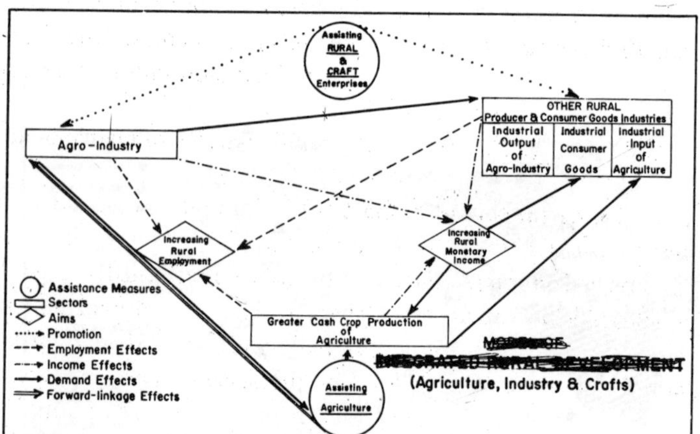

FIG 12.2: Model of Integrated Rural Development

Suggested Remedies

In principle, when plans promoting rural industrialisation are formulated, due allowance should be given to checks on the assistance programme and the evaluation of targets achieved. For example it is useful to remember that –

(i) "Action Plan" for rural industrialisation of this district is *one important instrument* of rural development. It is interconnected or interlinked with other aspects of rural life. Crucial importance herein is attached to assisting agriculture and improving interregional communication. In this connection an increase in agricultural cash crop production on account of completion of Bhima's Ujjani Dam project is expected to be of great industrial value,

(ii) The *motivators*, i.e. the Government agencies *are not subject to any* incentives *on the functions they perform*. Nevertheless they have a rationale for their existence. On the other hand, they have a tendency to expand their sizes. It is, therefore, the obligation of the Government to see that "net benefits" are given to those for whom they have planned.

(iii) Increase in industrial output and productivity through optimal utilisation of resources is the only cure of Socio-economic ills in DPAP area. The practical way to crack and break the vicious circle lies in the adjustment of policy-makers and administrators with multi disciplinary teams to plan, to direct, to regulate, and to evaluate the progressive steps towards rural welfare.

NOTES

1. International Development Agency (Nov.14, 1974), *Appraisal of Drought Prone Areas Project in India.*
2. District Statistical Officer, Solapur (1979-80), *Socio Economic Review and District Statistical Abstract of Solapur.*
3. Bank of India (Lead Bank Scheme) : *Solapur District Credit Plan for 1981, 1982 and 1983.*

13

SANDSTONE-SLAB QUARRYING IN THE BIJOLIAN PLATEAU OF RAJASTHAN

R.L. HIRAN

HOUSE-BUILDING is a primary need and utilisation of locally available building material is generally a universal practice. But the Vindhyan sandstone has long been used even in far off places. The Upper Vindhyan formations provide excellent sandstone in the south-eastern and eastern parts of Rajasthan in the districts of Bharatpur, Jaipur, Sawai-Madhopur, Bundi, Kota, Jhalawar, Bhilwara and Chittorgarh in a continuous belt, and west of the Aravallis in Jodhpur and Bikaner districts in detached patches which are being used as building stones regionally. Palaces, forts, and mosques at Agra, Bharatpur, Fatehpur Sikri, Delhi, Lahore, etc. and buildings of many settlements in adjoining States have been constructed from the Vindhyan sandstones. M.S. Krishnan has rightly remarked, "the Vindhyan sandstones, especially of the Bhander Series, constitute a great storehouse of excellent stones which, because of their regular bedding, uniform grains, pleasing colours, easy workability and durability, have been very extensively used all over Northern India as building stone[1]", "in which the stone-cutter's art," remarks Dr. V.Ball, "is seen in the highest perfection".[2]

Sandstone-slab quarrying activity in the Bijolian plateau has newly been developed. Hence, the present paper attempts to appraise the extent of the role played by the Vindhyan sandstone-slab quarrying activity in the area development of that part of the Bijolian plateau which lies in the south-eastern part of Rajasthan between 25° and 25°30′ N latitudes and 75° and 75°45′ longitudes coinciding with the eastern part of Mandalgarh Tahsil of Bhilwara District, south-western part of Bundi Tahsil of Bundi District and northern periphery of Begun Tehsil of Chittorgarh District. The paper also analyses the location of quarrying villages in their physical frame and discusses causatively the pattern of demographic growth and economic development. Some suggestions have been put forward to minimise the problems faced and to speed up the economic development of the area.

Geology

Geologically, Bijolian plateau lies in the western part of the Vindhyan Province of Central India and on the eastern borders of the Aravalli chain. The Great Boundary Fault, which has a throw of 1525 m, passes along the north and north-eastern periphery of the plateau, where the undisturbed, almost horizontally bedded, Bhanders have been brought against the highly folded and foliated schists of the Aravallis. Wadia maintains that "this junction is not of the nature of an ordinary fracture or dislocation, but marks the approximate limit of deposition of the younger Vindhyan sandstone against the foot of the Aravallis which was modified subsequently by faulting and thrusting."[3]

But M.S. Krishnan mentions that due to compression they have been affected by overthrusts from the west, along the Main Boundary Fault "which testifies semi-contemporaneous earth movement[4]". The crustal movement succeeding the Vindhyan deposition was in the form of the epeirogenic upheaval which lifted up the Vindhyan deposits from the floor of the sea to form a continental land area. The Vindhyan deposits are about 1200 m thick in the Bijolian plateau.

As the margins of the Vindhyan basin show a good development of sandstones, the plateau, lying on the western margin, is also richly endowed with this building stone. The prevalence of current-bedding and ripple marks in the strata points to its origin in the shallow agitated water of the coast, near the mouths of rivers and the constantly changing velocity and direction of its currents, while the red sandstones, of the Kaimurs and Bhanders indicate semi-arid and continental conditions which is also suggested by the perfect rounding of quartz grains in the majority of sandstones.[5] These sandstones are fine-textured, soft and easily workable stones of a deep red tint sometimes with white specks passing through variegated light shades of pink, buff, cream, grey and white, occasionally showing red streaks. They can be easily quarried in any quantity of any size. It is for such an assemblage of characters that they have been rendered eminently suitable for building or architectural work. When thinly stratified, the rock yields flags and slabs for paving and roofing purposes and when coarsely bedded, large blocks and columns can be cut out of it.[6] In the Bijolian plateau, mainly slabs and flags are cut from the open-pit quarries.

Topography and Location of Quarrying Villages

The Bijolian Plateau, regionally better known as the Uparmal, is situated above the mal (plain) area and is clearly and abruptly bounded by 1400′ contour line in the southern and western parts, 1300′ contour in the north and 1000′ contour in the east and north-east. The plateau is higher in south-western part, bounded by 1800′ contour around Rawarda and 1700′ contour on the southern border, marking an abrupt steep slope. The eastern margins of the plateau are less steep and contours from 1000′ to 1500′ run roughly parallel, perhaps in a step-like manner, in the sandstone area of south-western Bundi Tehsil. One north-easternly narrow offshoot of the Vindhyan formations runs from Bundi town towards Indargarh which is also clearly bounded by 1000′ contour, rising sharply up to 1300′. The spot-heights at quarrying villages vary from 1906’ near Nayanagar in the south-west and 1683′ at Bijolian in the middle of the plateau to 1263′ at Kharipur in south-eastern part; although some trig-stations indicate the elevation of 2014′ near Mindki Mahadev and 1946′ near Tilasma Mahadev on the southern periphery of the plateau. The higher portions in the middle of the plateau are near Salawatia-Gopalpura-Shambhupura which serve as water-parting of the small streams, viz. the Eru Nadi flowing from about 1800′ eastward in the southern part of the plateau to meet the Chambal at 966′ height; the Mangli and other tributaries of the Mej flowing north and north-eastward to meet the Chambal; and the Menali flowing westward from 1800′ with a fall of 120′ at Menal and then north-westward to meet the Banas and the Berach at Triveni near Bigod.

A large number of sandstone-slab quarrying villages in operation are presently situated in the southern part of the plateau mainly in the vicinity of the Eru Nadi right from Aroli through Kasian and Gudha in Mandalgarh Tahsil to Lamba Khoh, Rajpura and Gudha in Bundi Tehsil and the upper reaches of the Menali with Rawarda, Menal, Nayanagar, etc. being sites wherein the streams have washed away subsoil canopy from the valley slopes.

The rocks are exposed to weathering and erosive processes, hence the plateau except on the borders is devoid of compact forest coverage. The thin veneer of soil in depressions and valley floors supports agricultural practices in such quarrying villages.

The whole plateau abounds in sandstones but in northern and north-eastern parts slab-cutting stones are found at slightly greater depth where the quarrying lease-holder would not like to bid for indeterminate and delayed profits. Hence, these areas are not in operation, where boring determination survey must precede extraction. Sandstone extraction activity is therefore, concentrated on the southern and eastern parts of the plateau where quality sandstones for slab-cutting are available. They are quarried from surface to a depth of about 21′ (7 m). Drainage seems to have played a major role in sculpturing the topography for easy access to underneath layers.

Quarrying Activity

Sandstone-slab quarrying activity in the Bijolian plateau developed in 1940's when Kansia quarries were brought under operation, although hitherto some stone-flags and slabs used to be cut from nearby villages and transported on bullock-carts and camels to meet local needs. Since the plateau is about 500′ above the adjoining plain area, the heavy head-load or animal haulage could hardly economically meet the growing demands of distant places. Truck-transportation came to the rescue and facilitated haulage of bulky and heavy loads of even large size slabs to greater distances in States adjoining Rajasthan. As demand increased, new sources of supply were tapped. Now slab-quarrying activity is performed on a large scale in villages shown on the map.

Good quality red sandstone slabs had been extracted from Kansia and Kesarpura quarries for about three-four decades, but now they have been virtually deserted after quarrying for about 15′ to 20′ depth due to non-availability of quality sandstone underneath. Presently, whereas best quality red sandstone is quarried at Patpadia, Gurha, Baglya, Adekhali, Lamba and Khoh villages in Bundi Tahsil and grey-white sandstone at Budhpura and Gopalpura in Bundi Tahsil, and Shyampura and Sukhpura villages in Mandalgarh Tahsil, mixed shades are available at Rawarda, Aroli, Jhareli and Berisal Niwas villages in Mandalgarh Tahsil. Depth of these working quarries varies from near surface to about 21′ (7m).

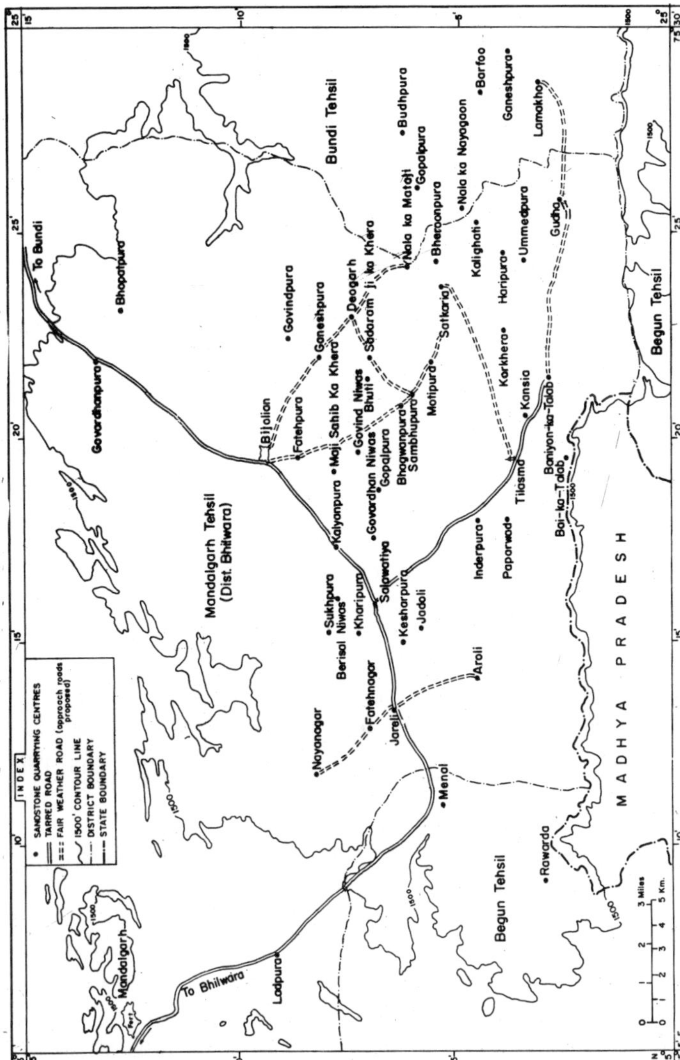

FIG 13.1: Bijolian Plateau : Sandstone Quarrying Centres

Following the development of quarrying activity on the plateau, the earlier Mandalgarh-Bundi fair-weather gravel road was metalled during the various five-year plans. Chittorgarh and Begun were connected with this road near Ladpura. One road has newly been extended from Salawatiya to Baniyon-ka-Talab, near the M.P. boarder. The area is poorly served by road network and paucity of at-least approach-roads is badly felt.

There are slight variations in the licencing pattern and rate of royalty in various mining (quarrying) circles of the State of Rajasthan. On the Bijolian plateau itself, while rent-cum-royalty leases are granted for a period of 5 years in adjacent areas of Bundi circle, its duration is only one year in Bijolian circle. Number of such annual leases fluctuated from 3814 in 1972-73 to 2096 in 1974-75 and again to 3085 in 1976-77 in Bijolian circle. Rate of royalty per tonne of dimensional stone varies from Rs. 7.00 in Bharatpur and Rs. 6.50 in Sawai Madhopur Districts to Rs. 5.00 in Kota, Bundi and Jhalawar Districts and Rs. 4.00 in other districts of Rajasthan. But there are marked differences in the cost of production and selling prices. Although the approved Govt. rates for labour charges are Rs.1.20 per running foot of roofing slab and Rs. 40 per 100 sq. feet of flooring slabs, the labourers are paid by the licence-holders Rs. 3 per roofing slab of any length (generally from 8′ to 12′) and Rs. 30 per 100 sq. feet of flooring slabs. By adding royalty, 8 per cent sales-tax and transportation-charges, these slabs are sold at Bhilwara, at a distance of about 100 km from the supply area, at the rate of Rs. 5 per running foot of roofing slab and Rs. 130 to 135 per 100 sq. feet of flooring slab. The brokers and suppliers who also own their trucks get sufficient commission and profit. Since the labourers are not adequately paid in keeping with their manual labour, their economic condition is very poor. Intoxicants, particularly country liquor, taken to withstand manual labour and to relieve tiredness, consumes a major share of their hard-earned wages. Other related social vices are also prevalent on the quarrying areas. Hence, check on excessive consumption of liquor is needed. It may be done through regulated supply of such intoxicants.

The labour castes are mainly the Bhil (the Mamas), the Chamars and the Rajputs who come from hilly and tribal areas, viz. Asind-Badhor-Beawar area, in Kota, Chittorgarh and Sikar Districts of Rajasthan. They reside in the thatched huts constructed by the contractors in the vicinity of the quarries. They become pets of their 'masters'.

The quarrying activity has generated the process of area development. Besides agricultural operation, mainly the kharif-cultivation, quarrying is the second principal occupation on the plateau. The immigrant labourers, contractors, suppliers, other officials, etc have increased the population. The income from stone-business and increasing circulation of money have attracted the retailers and stall-operators. Tea-stalls *dhabas* and hotels for lodging and boarding purposes have been established at quarrysites and along the roads, particularly on the *chaurahas* (cross-roads). A large number of such villages grew in size, and the population has more than doubled during 1961-71.

The quarrying activity on the plateau has huge potential and if properly handled can cater to the demands from distant urban centres of India. Sandstone slabs are not only used in the local Districts of Bhilwara, Bundi and Chittorgarh but these are also supplied to Jaipur and Sikar Districts in the north and Udaipur and Banswara Districts in the south in Rajasthan; Singoli, Neemuch, Ratlam and Indore areas in M.P.; nearly whole of Saurashtra up to Somnath and Surat and Vadodara in Gujarat and Bombay in Maharashtra. In distant areas, generally urban centres have great demand for sandstone slabs quarried on the Bijolian plateau.

Suggestions

In order to further increase quarrying activity on the plateau, problems faced by the people must be removed or minimised in the light of the following suggestions:

(i) Drinking water in drums on bullock-carts is brought to the quarries from a distance of 3-4 km where the wells show a low watertable of about 40′-50′ depth. So, the Government should make efforts for regulated supply of water at quarry-sites through water-tankers; make borings in the valleys and beds of the Eru and other streams for hand-pump water supply and construct small reservoirs to make use of water and keep the water-table from sinking,

(ii) The Department of Mines and Geology, Govt. of Rajasthan, should conduct sample block-cutting surveys in the northern part of the plateau to ascertain the depth where slab-stone may be available. If it is available at greater depth, either the

Government should remove the upper layer and then give on lease or may give the area on concessional leases,

(iii) Construction of approach roads to the quarries for round the year activity, and

(iv) Proposal for constructing the railway line between Kota and Chittorgarh, should be taken up by Government of India for quicker development of the plateau.

If these problems are urgently looked into the life of the rural people may take a turn for the better.

NOTES

1. Krishnan, M.S., *Geology of India and Burma*, 1968, p. 194.
2. Wadia, D.N., *Geology of India*, 1975, p. 427.
3. Ibid., p. 127.
4. Ibid., pp. 186-87.
5. Ibid., p. 426.
6. Ibid., p. 127.

14

ROLE OF CO-OPERATION IN RURAL DEVELOPMENT OF GUJARAT

C.S. RATHOD

THIS PAPER is devoted to the study of the role of various types of co-operative societies and banks in the rural development of Gujarat. How geographical factors influence such development and how knowledge of geography can be fruitfully utilised have also been examined.

The aim of 'Co-operation' is to promote better living, better business and better methods of production among the members and the community. Large number of co-operative societies are working in the rural areas of India aiming at providing credit to farmers for production and investment purposes. There were 95,782 primary agricultural co-operative credit societies in 1978-79 operating in 5,65,625 out of 5,85,004 villages, serving the rural population of 28 States/Union territories of the country. These societies are the pivot of rural development providing credit to farmers, including small farmers, agricultural labourers, rural artisans and others. Besides credit, rural people benefit from primary marketing and processing societies, and different types of cooperative farming societies — Farmers' Service Societies and large-sized Agricultural Multi-purpose Societies, Cotton Ginning and Pressing Societies, Sugar factories, and Weavers and other industrial Societies. These have greatly helped rural producers by providing credit, marketing and processing their produce, providing them employment, building schools and health centres in rural areas, offering better prices for their produce, constructing rural roads and controlling the powers of merchants, millers and money-lenders. These show how co-operatives promote rural re-construction. Long back, 'Rural Progress Through Co-operation', a UNESCO Publication, greatly admired the role of co-operation in rural progress.

Gujarat is the top-ranking state of India in the field of co-operation. It has done remarkable work in areas such as rural credit, rural and co-operative marketing and processing, co-operative dairies and sugar co-operative factories, forest labourers' societies and weavers' societies. It has been the accepted opinion that rural development of Gujarat is attributed to the fine working of co-operatives. The work done by AMUL (Anand Milk Union Ltd.) in Kaira District is well-known and its example is not only copied in Gujarat but also in whole of India and other under-developed countries. AMUL has changed the geography of villages of Kaira district — farmers get employment, better prices for their products, and agricultural credit; rural roads have come up along with agricultural and dairy college; schools and colleges; women of farmers are looking after animal husbandry programmes; young boys and girls learn in Youth Co-operative Programmes, how to take up co-operative leadership; backward classes and small farmers live in better houses and Sardar Patel University is devoting valuable academic resources in studying the various aspects of rural development. AMUL Dairy has made Anand town a tourist centre, a place of pilgrimage to those who want to study co-operation and co-operative dairies. The geographical reasons for such laudable works are : fertile soil, good rainfall, absence of flood diligent farmers, and devoted leadership. In contrast to this, West Bengal, Bihar, Assam, Orissa, Rajasthan, Jammu and Kashmir are cooperatively backward and poor in rural development because of various natural and human factors. A time has come for specialists of geography to study and suggest solutions for such situations.

To carry the fruits of co-operation and rural development among the weaker sections of the rural community and backward areas of Gujarat, there are agencies such as Small Farmers' Development Agencies, Drought-Prone Area Programmes, Regional Rural Banks, Command Area Development Programmes, etc. In all these schemes in Gujarat, Co-operative agencies are actively associated and make a significant contribution. To take benefits under the SFDA, for example, farmers are first enrolled as members of Co-operative Credit Societies and then they become beneficiaries under schemes such as minor irrigation, cattle, poultry, improved agriculture and other activities such as — piggery, fisheries, goat- and sheep-rearing, bee-keeping, sericulture, duck-rearing etc. Over and above this, to broadbase the benefits of Co-operation to the neglected sections, on the recommendation of the National Commission on Agriculture, Government have formulated special schemes such as Farmers' Service Societies (FSS) and large-sized Agricultural Multi-purpose Societies (LAMPS) exclusively for the benefit of these persons. In Gujarat the number of FSS was 40, and LAMP 58 as on June 30, 1979.

In the days to come, specialists in geography will have to play a greater and crucial role in this area:

(1) Experts in economic and commercial geography should take interest in the co-operative field and highlight in their studies and research importance of co-operative dairies, co-operative sugar factories, cotton ginning and pressing mills and names of products like AMUL and NOGA.

(2) Generally uneducated rural people do not understand figures and statistics which are cold and unimpressive. Instead, by drawing educative and understandable maps, a domain of geography, geographers can demonstrate the benefits of 'Co-operation' to villagers.

(3) Geography is the arithmetic of human welfare and human life. Co-operative form of organisation has promoted new type of Social geography — work together, live together, 'one for all and all for one'.

(4) Because of adverse topography, long distance and inaccessibility of vast areas of the country, rural progress and performance of institutions have been unfavourably affected; this point was revealed by RBI in its study on Regional Rural Banks. This is an area where the geographer can throw some light.

In short, geographers of the country who have cartography as their basic tool will have to use their talent and skill in the co-operative field, which is an important agency of rural change, rural development, rural regeneration.

PART IV
ROLE OF RURAL TRANSPORT AND MARKETING

15

RURAL MARKETS – A Perspective for Development (A Case Study of Rajasthan)

H.M. SAXENA

ECONOMIC DEVELOPMENT is a long-term process that occurs over decades and generations, therefore, there is an urgent need to broadbase our development planning for the masses. In our country, although some attention has been given to rural development, it still requires greater emphasis and a new perspective. Among the various approaches of rural development, an important one is the development of a network of rural markets to serve both as markets and as nuclei of development.

What is the need for more rural markets? A question may arise, when there already exists a network of markets. Markets are many, no doubt, but their spatial pattern is such that they serve a limited space and are mostly located in towns. Thus they not only create regional imbalances but are also not approachable to rural masses. Johnson has rightly emphasised that "every rural producer should be within convenient travel time to some adequately competitive selling place for his produce, some equally competitive sources of consumers' and producers' goods and some adequately diversified service centre".[1] But he qualifies this statement by saying that the market centres are only the basic building blocks of a much more complex hierarchy of central places, capable of knitting the entire spatial economic structure together in a functional sense.

The main objective of the present paper is to study the rural market forms as these exist today in the State of Rajasthan and also to provide a perspective for their development planning.

Rural Markets in Rajasthan

The rural markets and marketing system in the State of Rajasthan are still rudimentary, unsystematic and diversified in nature. The goods and services of permanent markets are not available to a majority of the rural consumers, because people mostly live in villages far removed from towns. The diversified nature of geo-economic and cultural conditions are responsible for varied pattern or forms of rural markets in the State. The various forms of rural markets and marketing as identified in the region under study are as follows:

(i) Periodic markets;
(ii) Mobile traders; and
(iii) Rural shops or permanent rural markets.

(i) Periodic Markets

The existence of periodic markets is a characteristic feature of rural marketing in Rajasthan. This is because the state of agriculture is subsistence and every peasant has something to dispose of and has his limited wants to satisfy through periodic markets. Two such periodic marketing institutions that have developed in the state are 'weekly markets' and 'fairs'. The difference between a weekly market and a fair lies in the fact that a weekly market is primarily a weekly assemblage of local traders organised to facilitate retail operations. The fairs on the other hand occur occasionally, may be once or twice a year. Both these trade institutions require special mention in order to understand the rural marketing system.

Weekly Markets: Although weekly markets are associated with peasant agrarian economy, they are not held in all parts of the State. These are prevalent in south-eastern and some eastern and central parts of the State. Out of a total of 26 districts only 6 districts, viz. Jhalawar, Kota, Bundi, Tonk, Sawai-Madhopur and Jaipur, have got such markets. All these districts are having agriculture as the basis of livelihood and economy and are also characterised by level relief, either plain or low plateau. However, the existence of weekly markets in this region is more due to historical and cultural factors, although geographical factors also play an important role in selection of site, schedule, layout of markets, etc. These markets were mostly started by local Chiefs in order to provide a platform of exchange of commodities between producers and consumers and also provide a place for social contact. After initial success these markets continued to grow and still exist. They now serve not only the rural but also the urban population.

In the region mentioned above weekly markets are held once in a week and are locally known as *hat*. These markets generally follow a weekly cycle and resemble more or less Stine's model, which accounts for a periodicity of markets and mobile behaviour of the merchants.[2] The mobile merchants adjust themselves by visiting several markets on a regular basis. They have an additional advantage of concentrating demand in specific markets on specific days.

In these markets people flock in large number from 10 a.m. to 3 p.m. depending upon the climatic conditions. The peak-hour of business is noted about 3.30 p.m. while the villagers start dispersing from 4 p.m. onwards. The arrangements in the weekly markets is orderly, sections are earmarked and many traders have their fixed places.

These markets are a place for the sale and purchase of different types of commodities, produced either locally or assembled from outside. Thus a weekly market serves as a place where rural people bring their seasonal agricultural products for sale and make purchases of daily consumable articles, artisan products, etc.

Fairs: Fairs rank among the oldest as well as the most fascinating of trading institutions. Even today, fairs are an integral part and play a vital role in the rural economy of Rajasthan, because, in a fair, assembly of people creates opportunity for retail trade and they are most suitable to an agrarian economy for trade in specialised products.

Unlike weekly markets, fairs are held throughout the State irrespective of their geographical conditions. The total number of fairs held in Rajasthan is about 1500. But out of total nearly 1200 are one-day fairs associated with festivals. These fairs are simply held only for a few hours in the noon or after-noon and have very little economic importance. On the other hand, there are economic fairs dealing in buying and selling of cattle, commodities and articles of daily use. Trade of cattle is a special feature of as many as 65 fairs of the State. Among cattle fairs 9 fairs are prominent, which are of state level and organised by State Animal Husbandry Department.

Although, fairs are held throughout the year in some part or other of the State, it has been observed that these are more frequent during the off-ploughing season. In fact these fairs provide a marketing occasion for the rural inhabitants and are a powerful medium for economic, cultural, social and religious intercourse of the people in the region.

(ii) Mobile Traders

Mobile traders are those salesmen who visit rural areas in an individual capacity and on the days which they like best. This practice has been developed because these areas are having "very low density of demand and/or high friction of distance".[3] Such traders, although common throughout the State, are more regular and frequent in districts of Ganganagar, Churu, Jhunjhunu, Sikar, Alwar and Bharatpur, while in Ajmer, Bhilwara, Udaipur, Chittorgarh, Dungarpur, Banswara and Sirohi their mobility is restricted by rough terrain and they visit rural surroundings located near cities. On the other hand their frequency and number of visits are minimum in arid western part of the State. These mobile traders provide articles of daily use to rural people and sometimes also purchase agricultural products from villagers in limited quantity and also on cheap rates. There is a complete absence of such traders in areas served by weekly markets.

(iii) Village Shops

In villages, particularly in those villages which are having more than 2000 population, some permanent shops have come up which serve as bases for trade. Mostly, these shops deal in grocery items, cloth etc. while some function as service-cum-sale shops. No special arrangements have been made for these shops and generally a small room facing the road is used for display and sale of commodities. With the growth of settlement and population such shops do take on new dimensions and sometimes provide a base for market development.

Perspective for Development

The present system of rural markets, as discussed above indicates that varied forms of rural markets are prevalent in different parts of the State. Therefore, a uniform policy of development would not be effective in all parts of the State. A planned development of rural markets is only possible through a proper spatial system with different priorities fixed for development of various regions.

Thus, on the basis of the present structure of rural markets and for their further planned development, Rajasthan State can be divided into the following three regions:

(i) Region of weekly markets;
(ii) Region of mobile traders; and
(iii) Region without rural trading facilities.

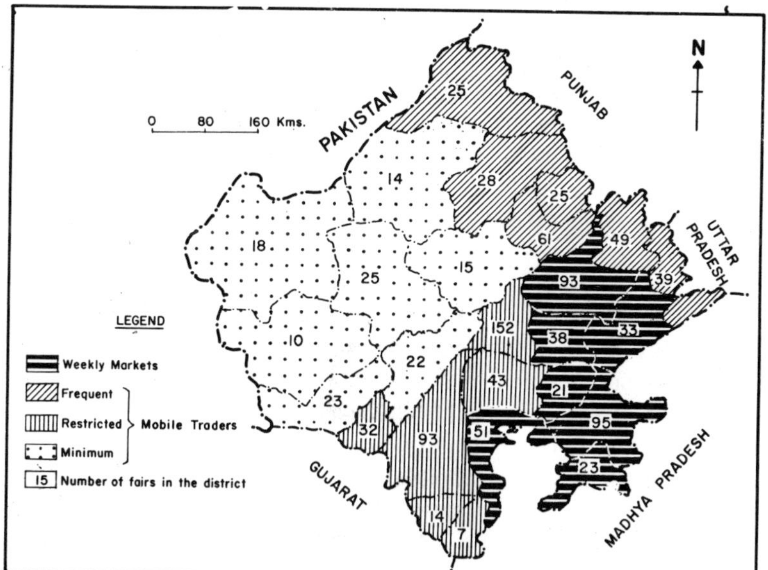

FIG 15.1: Rajasthan : Rural market pattern

(i) Region of Weekly Markets

In south-eastern and central parts of Rajasthan, where weekly markets are common, there is a tendency towards some stablisation of trade activity as well as development of permanent shops. This is also due to the growth of population of the villages and also due to economic growth. Therefore, there is a need for selection of suitable sites for rural markets. Care must be taken that their number should not be many. They must correspond to regional patterns of ecology, transport, population and potentialities of economic growth, etc. These market centres may be developed in such a manner that they must be surrounded by clusters of villages and should not be so near the big cities that these might be totally eclipsed.

(ii) Regions of Mobile Traders

This region comprises northern, north-eastern, southern and south-central parts of Rajasthan. In this region rural marketing has not developed, either due to lack of incentive or due to adverse geographical conditions. But at present many districts of this region, viz., Ganganagar, Alwar, Bharatpur, Ajmer, Bhilwara, Chittorgarh have developed the agricultural base of their economy. This indicates possibilities of the development of weekly markets as well as of permanent rural shops in bigger settlements. A proper integrated system of weekly markets will definitely provide an extra impetus for economic development.

(iii) Region Without Rural Trading Facilities

There is a complete absence of rural marketing in the areas of arid western Rajasthan due to high density of sand-dunes, scarcity of water, low density of people and other unfavourable geographical conditions. Only fairs are common in this region but these too are limited in number. This region is neither favourable for mobile trades nor for weekly markets. The only possible alternative in this region is a planned development of mobile traders with fixed routes. Such traders should be provided with extra incentives.

The present study is just a preliminary survey of rural markets but attempts to provide an insight into rural development programmes. A State level agency must be established for conducting surveys preliminary to developing a rational system of marketing centres, and to provide finances and other infrastructural facilities, etc for the same. The establishment of rural market system tends to make agricultural sector increasingly market sensitive and also provide an incentive for village producers to increase production. This works as a catalyst to the rural development process.

NOTES

1. Johnson, E.A.J., *The Organisation of Space in Developing Countries*, Cambridge, Mass, 1970, p. 418.
2. Garner, B.J., "Models of Urban Geography and Settlement Location," *Models in Geography*, edited by Chorley, R.L., & Haggett, P., Methuen, 1967, p 320.
3. Bromley, R.J., "Periodic Markets and Rural Development Policy," *Cambria*, vol. 2, No. 2, 1975, p 156.

16

TRANSPORT AND MARKETING FACILITIES IN A BACKWARD REGION

A Case Study of the Patha Area of Bundelkhand

R.C. SRIVASTAVA AND JAWWAD ALI

BACKWARD AND economically depressed regions develop a weak spatial organisation on account of uneven economic topography particularly in the absence of sound network of transport and market centres. Due to wide gaps between rural areas and market centres on one hand and poor accessibility to transport routes on the other, backward regions suffer a lot in the free flow of inputs and regular disposal of produce. As transport lines establish relations and make for integration among all the economic activities over space, an efficient transport network is vital for the rejuvenation of rural economy.[1] Similarly the development of appropriate number of market centres is essential as they act as service-and-growth-points to generate significant growth impulses and exert influence on their surroundings for economic development.[2] Existing market centres and their facilities for our rural folk are quite small and insufficient. By this time, if there were to be an emergence of proper distribution of market centres with necessary infrastructure on any given rural landscape, indeed, they could have acted as energisers, for the transformation of rural society in the sphere of socio-economic change.

Study Area

The Patha area which lies to the south of the first Vindhyan Range in the District of Banda (U.P.), is the most backward area of

Bundelkhand. Extending in an area of 1149.55 sq.km, it includes 107 inhabited villages of Karwi and Mau tehsils. It is a sharply defined upland plateau of some continuous ranges and isolated hills. It is a drought-prone and a water-scarcity area because of its lithology, physiography and climatic conditions. Agriculture is the backbone of the economy as it provides employment for about 88 per cent of the total workforce. Livestock rearing is mostly practised as a supplement to agriculture on a mixed farming basis in the whole area.[3] Industrial sector plays a minor role as only 2 per cent of the working population is engaged in household industries and manufacturing.

Patha (68,197 persons in 1971) is a rural area with about 89 per cent of its population living in the villages. The density of population is very low (66 persons/sq.km). Small villages (200-499 persons) constitute 42.3 per cent of the total villages of the area while large size villages (above 2000 persons) represent only 5.6 per cent. The spatial and functional integration of settlements and the economy is really a problem in the area due to wide spacing of settlements (above 1.90 km) and poor service facilities which are unevenly distributed over the physical space. The poor rural-urban interactions, absence of periodic markets, low density of roads and poor communication linkages have made Patha an area of depressed economy.[4] Manikpur is the only urban centre which provides the major part of social and economic services to the area.

Objectives

In the light of overall condition of this backward area the main objectives of the present study are: (i) to analyse the transport network by computing density, accessibility and degree of connectivity of roads, (ii) to explain the spatio-temporal distribution of markets, (iii) to identify farmer-seller travel patterns, (iv) to highlight the problems of transport and marketing facilities, and (v) to suggest a sound network of roads and market centres for improving the present defective transport and marketing systems and to obtain an optimal spatial organisation for integrated development of this area.

Methodology

In the study of the transport system and network analysis secondary data have been availed from records and administrative reports of State and Central Government offices. The study of the present system and problems of marketing faced in the area is based on primary data which have been collected with the help of household as well as village schedules covering all the inhabited villages. In order to find out space preference for marketing 10 per cent of the households of each and every village were randomly selected. Questions were asked about the mode of sale of agricultural and animal produce, type of marketing agencies and channels present, preference for market centres and problems faced in the disposal of surplus produce and products. Transport network and farmer-seller travel patterns have been analysed by applying suitable quantitative and cartographic techniques.

Transportation Structure

A. Modal System

Transport arteries are a more real feature of the geography of an area than terrain from the view point of human action.[5] The area enjoys the services of two railway lines, namely, Allahabad-Jabalpur and Manikpur-Jhansi (Fig.16.1). Their total length in the study area comes to 95 km. The area served by these lines is small but the linkage relationship is significant as the former links M.P. and Eastern U.P. and the later Bundelkhand region and Delhi. Railways are the main transport routes for inter-regional trade dealing mainly with forest products.

Road system of this area is not inter-woven (Fig. 16.1), but the general framework of roads presents some typical features. Karwi-Manikpur road (13 km) runs parallel to the railway line while the next Banda-Allahabad road (14 km) passes through its northeastern part providing a branch (5 km) from Bargarh diversion to Bargarh market. Total lengths of metalled and unmetalled roads are 33 and 47 kilometres respectively. If the unmetalled roads are taken into consideration some transport nodes can be marked. The network of single path is common. Tree ordering system is absent. The regional patterns of roads indicate the geographical characteristic of the area concerned.[6] As the area is very small the regional patterns of roads are not clear. Vindhyan ranges and hills serve as static barriers to the development of roads. Forest roads have been developed with a good number of nodes and circuits (Fig. 16.1) but they are used only by the forest department. The network of dusty zig-zag cart tracks, interconnecting villages, is the most important constituent of the transport system in the region. As the people mostly follow the cart tracks for their short time journey a good number of villages of Manikpur Patha have been connected to the railway stations.

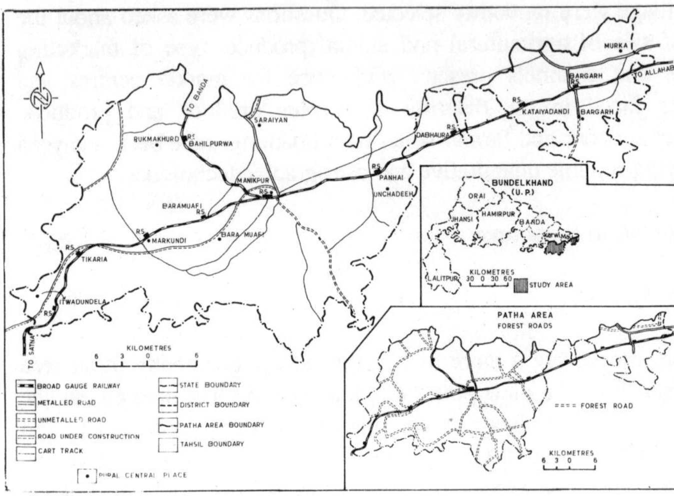

FIG 16.1: **Patha Area : Transport Network-1981**

B. Connectivity

The measurement of the degree of connectivity helps in finding out the levels of transport development in a region. The efficiency of the road network is better evaluated with the help of topological indices.[7] In the present study the 'gamma index'* has been taken into consideration for measuring the degree of connectivity. Table 16.1 shows the degree of connectivity of the whole Patha area and its two different regions.

It is clear from this table that Bargarh Patha has higher degree of

* The equation of gamma index (Y):

$$y = \frac{e}{e\max} = \frac{e}{3(V-2)}$$

Where e = Actual edges; max = Maximum edges; v = Vertices connectivity than Manikpur area because of better terrain and economic health conditions.

TABLE 16.1: Degree of Connectivity

	Vertices	Nodes	Degree of connectivity
1. Patha area	7	10	.29 or 29%
2. Manikpur Patha	4	6	.33 or 33%
3. Bargarh Patha	3	4	.50 or 50%

C. *Density of Roads*

The relationship between transportation and population is discussed and is used as the basis for examination of such additional factors as the physical environment, rail competition, intermediate location, and commercialisation.[8] The density of roads has been measured in terms of both the population and area. As the length of metalled roads is small the unmetalled roads have been included while finding out the density of roads. The table below reveals the variation in the road densities in both tracts.

TABLE 16.2: Density of Roads

Area/sub-area	Length (in km)	Length in km. per 100 sq.km	Length in km per 1000 persons
1. Manikpur Patha	60	8.5	15.3
2. Bargarh Patha	20	14.7	15.5
3. The Patha area	80	9.0	15.4

It is clear from this table that the density per 100 sq.km is higher than in other parts of Bundelkhand region where it ranges between 5 to 9 km/100 sq. km. Bargarh area is in better position than Manikpur area only on account of its terrain and location. As regards the density per 10,000 persons the sub-areas and the Patha area itself represent more or less equal density. In Manikpur area both type of densities are low on account of dissected plateau surface, dense forest patches and wide spacing of settlements.

D. *Accessibility*

In the present study the areas lying beyond 7 km of a road or railway station have been considered as inaccessible, although the true distance would, of course, be influenced by many factors like individual preference, and the availability and cost of alternative means of transport.[9] One to Seven kilometer isochrones have been drawn along the roads and around the railway stations to show the degree of physical accessibility (Fig. 16.2). The physical accessibility in relation to the rural settlements of the area has been computed as it is clear from the Table 16.3.

TABLE 16.3: Distribution of Settlements and Zones of Accessibility

Range of accessibility (km)	Accessibility to road No. of settlement falling per range	Accessibility to rail No. of settlement falling per range
0 - 1	11(10.58)	3(2.88)
1 - 2	10(9.62)	15(14.42)
2 - 3	7(6.73)	20(19.29)
3 - 4	7(6.73)	15(14.42)
4 - 5	11(10.58)	11(10.58)
5 - 6	8(7.69)	11(10.58)
6 - 7	4(3.85)	8(7.69)
Above 7	46(44.23)	21(20.19)

Note: Figures in brackets show the percentage of villages to the total number of villages.

The map showing accessibility to railway stations reveals that south-western and north-eastern parts of Manikpur Patha and southern part of Bargarh Patha are the areas of very low accessibility, (Fig. 16.2). 20.19 per cent villages are situated beyond 7 km while 17.31 per cent villages have easy access, i.e. between 0-2 km from railway station. The accessibility pattern of road is quite different from that of rail's accessibility. Due to mountain ranges, rugged terrain and wide expanse of forests, the major part of the Patha area represents very low accessibility to metalled roads (Fig. 16.2). Only 10.58 per cent of the villages are situated within 1 km distance. A large number of villages (44.23%) are inaccessible from the metalled roads. On the whole 14.42 per cent of villages are inaccessible by road and rail both.

E. *Functional Interaction*

The functional interaction in the central place system is best manifested in the road traffic flow.[10] The interaction between the different parts of the area is very poor. For example, the two main service centres, i.e. Manikpur and Bargarh, are not linked by road. Traffic flow by bus is not significant as only these two centres are linked by three metalled roads of the district. Manikpur has better functional interaction than other central places of the area as it serves as a growth point in this area. The bus frequency between Karwi and Manikpur is not appreciable as only ten buses ply up and down and out of these one bus goes to Banda and Rajapur separately. Allahabad-Banda road, which passes through the northern part of Bargarh Patha, has a frequency of 13 buses.

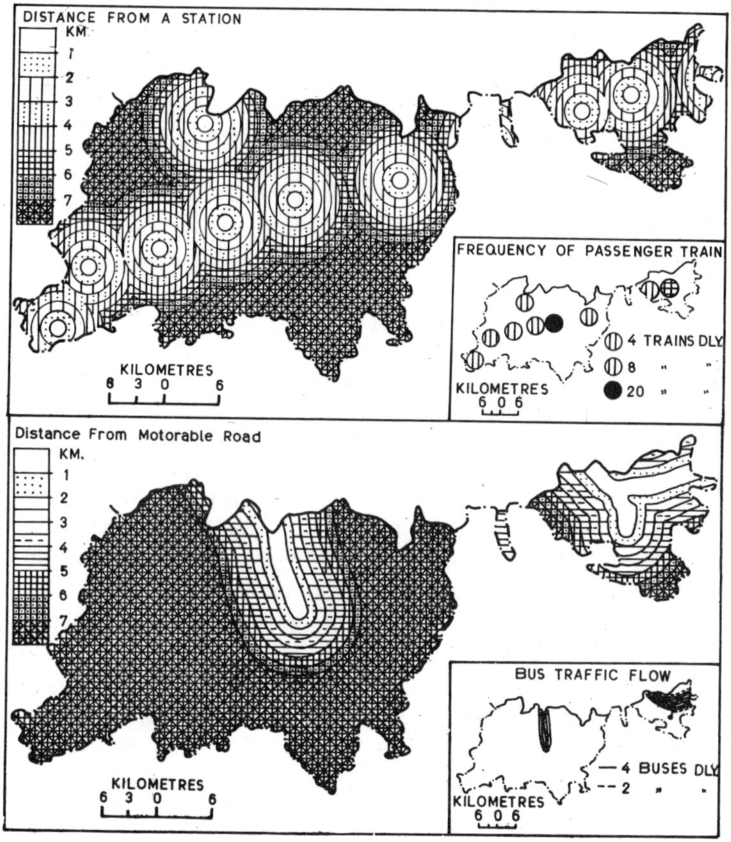

FIG 16.2: Patha Area : Accessibility by Rail and Road

This area is well linked by railways to different parts of the country. Six up and down trains run on the Allahabad-Jabalpur lines while the frequency on Manikpur-Jhansi and Manikpur-Lucknow lines is of three and one train respectively. Frequency of train stoppage at a station is more important than the number of trains running from that station. Manikpur and Bargarh have a frequency of twenty and eight train stoppage while the rest of the stations have the minimum frequency, i.e. four. Thus the frequency of passenger trains at many stations indicate deficiency in train services.

Marketing Structure and Facilities

In the regions of backward economy market centres, as service and growth points, exert influence on the surrounding environment for catalysing economic development. In the absence of daily market centres *hat* (periodic market) acts as a node of spatial interaction. In the rural areas the periodic markets not only serve as outlets for surplus local agricultural produce and shopping centres for the daily needs of the people but also function as suitable bases for social and agricultural contacts.[11] Periodic markets and market villages portray the real nature of spatial interaction between central places and their hinterlands.

Spatio-temporal Distribution of Market Centres

Market centres emerged only after goods began to move to consumers in exchange system transcending the limits of householding unit or the immediate social groups.[12] So long as the resulting exchange of goods was sporadic market centres did not develop. The major part of the study area in the 19th century was covered by dense forests where the *Kols* dominated for a long period. The exchange of goods was prevalent under barter system in the traditional society. Manikpur could grow as a market centre only after the construction of Allahabad-Jabalpur rail line in 1889. Bargarh, the second market centre, attained the status of a market in the early decade of this century. The views of Polyani and Pirenne that "markets do not develop out of the demands of purely local and individual commerce, but are primarily induced by external exchanges of complementary products with an alien population", seems to be true in the present context as not even a single hat (periodic market) could develop in an area of 1150 sq.km inhabited by 68,197 persons (1971 Census).[13] Because of agricultural backwardness, constant exploitation by *Zamindars* and contractors, the area developed an internal economy where permanent markets could hardly develop.

The two market centres, i.e. Manikpur and Bargarh, act as growth points in the two different parts of the Patha area. Manikpur (6512 in 1971), which is a block headquarter and an important rail junction of Bundelkhand, is a developed daily market centre. This centre caters to the daily and periodic needs of its own population and that of the surrounding areas. Besides, the retail shops of grain, cloth, hosiery, grocery, footwear, electrical goods, utensils, medicines and stationery, this centre also deals in wholesale of grains and cloth. For high order shopping goods majority of consumers of its surrounding areas visit this market. Bargarh (2378 in 1971) is the next daily market centre which caters to the needs particularly of daily and periodic goods, of 41 villages. It also enjoys the facilities of retail shops of cloth, hosiery, *kirana*, and grocery. Markundi is developing as a third daily market centre in Manikpur Patha area. Some retail shops of cloth, *kirana* and grocery have come up here. Besides these three daily market centres, some of the large villages have a few retail shops which deal in grocery items, sweets and cosmetics. Forty-two per cent of the villages have grocery shops while only two villages have retail shops of cloth.

System of Marketing

The system of marketing in a backward agrarian area differs from that in a developed area where the marketable surplus is high, transportation net is thick and other infrastructural facilities are good. Marketable surplus in agriculture is the very foundation on which the whole structure of our development stands. Among the marketable surplus items wheat, maize, millets and gram as the agricultural produce and milk and *khoa* as the animal produce, are significant in the study area. Products of household industries are insignificant and they do not find a place in marketable surplus. It is difficult to calculate marketable surplus of agricultural crops because production year varies from crop to crop. Thus marketable surplus is not a free variant but a forced surplus.[14] In the study area, as the majority of landholders are economically weak, the substantial holders who have enough marketable surplus, are the old Zamindars. So far as the food grains are concerned the marketable surplus entering the markets of the area is very low. This is because (i) low production per hectare, (ii) the increasing propensity to hoard, (iii) consumption of better cereals in place of inferior careals, (iv) poor road network, and (v) physical barriers, such as mountain ranges and dense forests.

Since there are no periodic markets in the Patha area the question of gathering of farmer-sellers in the rural areas does not arise. But for the sale of surplus agricultural and animal produce the farmer-sellers become bound to visit daily market centres located within, as well as outside of the Patha area. The usual channel for marketing of agricultural produce, particularly in the study area, is usually marked in three forms: (i) producer – village shopkeeper – wholesaler – retailer – consumer channel, (ii) producer – itinerant merchant – wholesaler – consumer channel, and (iii) producer – primary wholesaler – secondary wholesaler – retailer – consumer channel. For the last five years the rich farmers are selling surplus wheat to government procurement agencies through which the State Government purchases wheat in bulk in primary wholesale markets and sells at fixed rate through fair price shops in the towns.

Major quantity of marketable surplus, especially grains, is sold by farmer-sellers in their respective villages at unfavourable prices than can be obtained in the wholesale market. This type of sale generally takes place when the cultivator is in debt. Majority of marginal farmers remain in the clutch of old Zamindars, *sahukars* and contractors who purchase their grains, whatever surplus is available, in their respective villages. From the field study it was calculated that farmers of 30 per cent villages do not market surplus grains out of their own village (Fig. 16.3). On the other hand, the itinerant traders or *beoparies* also go from village to village for collecting the agricultural produce from farmer-sellers at comparatively reduced price. In the Patha area the itinerant traders purchase surplus grains only from 15 villages as it is clear from Fig. 16.3. Village trader, who deals in his individual capacity, is the farmost purchaser of agricultural produce. He usually collects the produce from villages and brings it in the wholesale market. The poor farmer-sellers are rather deceived and exploited by the brokers and itinerant traders. Local landlords also sell their produce directly to the village traders or town dealers visiting the village markets. *Arhatias*, who collect grains and later on sell in the wholesale markets, advance loans to the village merchants and traders on the condition that the produce will be sold to them or through them.

Though a major portion of agricultural produce is disposed off at village level, some parts of the produce, thus finds a place in the markets or *mandis* (wholesale markets) which are located at distances of 5 to 20 km (even more) from the villages in the study area. In the Patha area there are only three centres, Manikpur, Bargarh and Markundi, which receive farmer-sellers of 19, 14, and 4 villages respectively (Fig. 16.3).

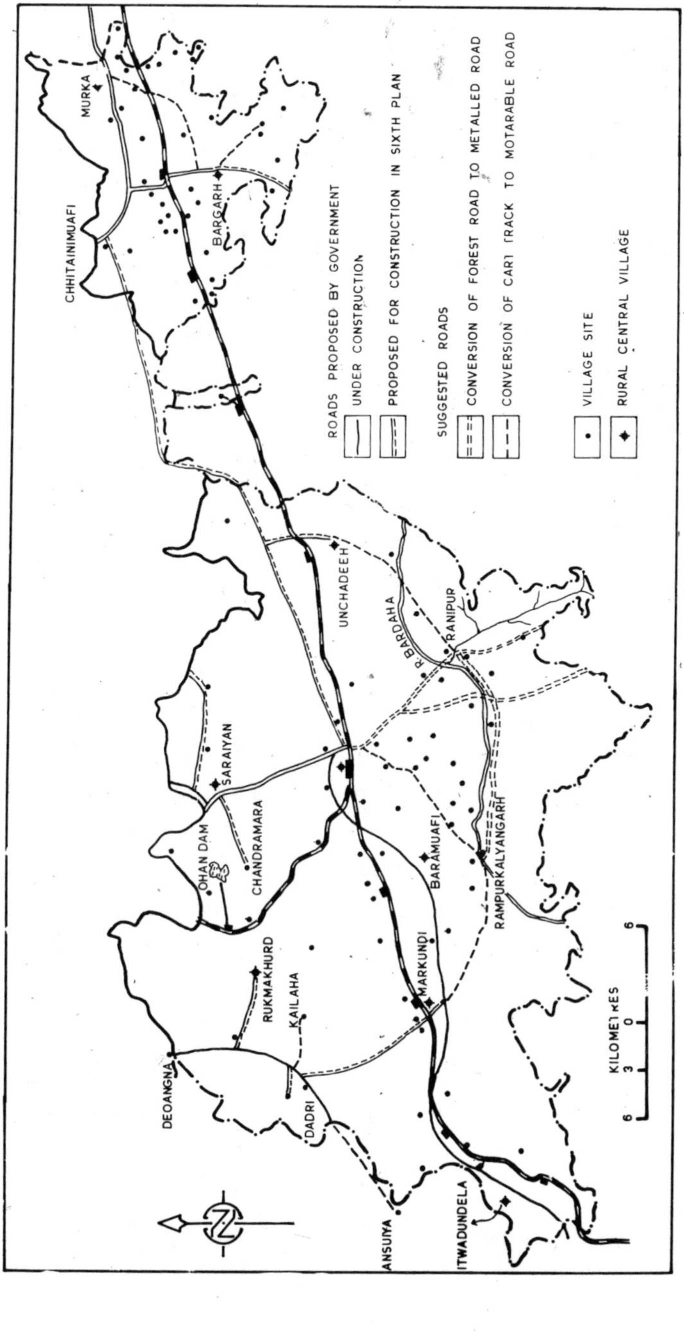

FIG 16.3: Patha Area : Proposed Road Network

Farmer-sellers of these villages sell their surplus grains directly in the markets. Karwi, an important town of Bundelkhand and a Tahsil headquarters, which is situated at a distance of 15 km from the northern border of the Patha area, receives farmer-sellers of 13 villages mostly of northern and north-western parts of the study area, whereas Majhgawan and Dabhaura, both situated along Allahabad-Jabalpur rail route, receive farmer-sellers of 4 and 1 villages respectively. The maximum distance covered by a farmer-seller reaches 20 km. Fig. 16.3 clearly shows with the help of visit lines the visit patterns of the farmer-sellers. For lack of information the circuit of itinerant traders could not be shown on the map.

The main animal produce in the study area are milk, *khoa*, curd, *ghee*, and skin but milk and khoa are the only marketable surplus items. Speedy assembling long distance transport and large scale processing being yet undeveloped in the country, the consumption of fluid milk is generally confined to the area of its production.[15] Though the production of milk is not encouraging yet the villagers supply surplus milk to nearby towns and service centres or sell to the middle men, who usually purchase milk in bulk quantity from different villages and sell in the towns at high price. There is no organised agency to purchase their surplus milk at reasonable price. The village level data reveal that 67 villages supply milk whereas some of the milkmen in 6 villages prepare *khoa* which they sell in Manikpur town (Fig. 16.3). The data also reveal that middle-men collect milk from 50 per cent villages and carry it by trains to Satna and Allahabad Districts where it is sold to retailers and consumers of towns situated along the rail routes. This intermediary system is a granite wall between the dairy plant and producers and the system exploits both of them.[16] With the help of space preference data the visit patterns of milk suppliers have been shown in Fig. 16.3 which shows that milk suppliers of 11 villages sell their surplus milk out side the study area, particularly in Karwi and Satna towns, while 31 villages are associated with Manikpur and Bargarh service centres.

Problems of Transportation and Marketing

The truncated spatial interlink between central places and the villages in their hinterlands have affected the entire fabric of national economy, for the impulses generated by new investments and innovations remained confined to a superficial urban layer being unable to percolate to the rural interior through the thwarted channels of circulation. Lack of better surfaced roads and bridges is the main problem facing the rural transportation. In our country, where 1 out of 3 villages is more than 5 miles away from an improved road, [17] the Gulf separating the rural masses from the national mainstream has widened. The genesis of the problem lies in the fact that rural settlement system established centuries ago in response to the then available transport technology, among other factors, has not readjusted to the fast changing transport technology.

In the Patha area the lay-out of the rail network was designed to facilitate military movement and to exploit the valuable forest products and minerals of the area. No serious attention was paid towards developing an integrated road transport network as it was a neglected area and remained a victim of the poisonous policy of regional disparities. With the result the vital road links in rural areas remain still largely neglected. The area is poor as regards the facilities of motorable roads because 35 km total length of metalled road is very low in proportion to its area and population. The roads link only two service centres while several emerging rural central places are still lacking this facility. Due to low accessibility to roads, particularly metalled road, majority of villages remain detached from development programmes. The movement of socio-economic and administrative facilities is largely affected due to the existing transport network. Major part of the area is under passive interaction on account of very low intensity of goods and traffic flow.

The hilly nature of terrain, reserved forest patches and numerous perennial and seasonal streams are responsible for the backwardness of the area in respect to transport. For the physical integration of settlements the existing road network is utterly inadequate. Moreover, in the present set up of rural transportation many villages are handicapped to avail the facilities of market centres for their development. It is thus observed that the meagre road transport facility is one of the basic problems of this area.

Communications from the field to the village and from village to the *mandi* (wholesale market) are often extremely poor and defective in backward areas. Bad roads and tracks connecting villages with the markets not only add to the loss of transportation and aggravate the strain on bullocks and other pack animals, but also lead to the multiplication of small dealers and intermediaries. They also restrict markets by cheap and rapid movement of agricultural and animal produce.[18] There is not only mileage deficiency but there is also a lack of feeder-roads connecting villages to market towns and nearest railway station so that produce cannot advantageously be transported to *mandis*. The villagers become victim of itinerant traders and local merchants and other intermediary agencies only on account of poor transport facilities.

The Patha area also lacks the optimal spatial integration and functional coordination of settlements due to absence of periodic markets. The major problem which is being faced in the absence of periodic markets is the improper and uneasy disposal of marketable surplus. Manikpur and Bargarh market centres handle a very low percentage of marketable surplus due to other market centres. Moreover, the absence of periodic markets compels the farmer-sellers to dispose of their surplus agricultural produce to local merchants and itinerant traders otherwise they will have to move to large market centres and *mandis* (located at greater distances) losing their time, energy and money.

The farmers of the study area are facing other problems of marketing which are associated with the present marketing channels and agencies. There is a lack of any kind of collective organisation among producers. Thus, it is common to find that producers of grains and milk as a class are being exploited by the purchasers. The farmer, in general, sells his produce at an unfavourable place and at the unfavourable time and usually gets very unfavourable terms.[19] Because of poverty and indebtedness, and the unsatisfactory nature of communication, the produce of this area is sold soon after the harvest when there is a glut in the market. The farmers who some times mortgage their standing crop on account of loan, do not get the price of the produce of the fields and sometimes they are forced to dispose of the major portion of their produce almost immediately after the harvest in order to pay the debt of the local merchant or land holder. There is a serious problem that superfluous middle men, who handle the major part of marketable surplus, have even the web in such a manner that the producer can hardly get himself free from the clutches of these intermediaries. Dr. Munshi found that in marketing the produce the cultivator gets somewhere between 42.3 to 73.7 per cent of the consumer's price and from 26.3 to 57.7 per cent goes to the middlemen.[20]

The marketable charges payable by the producers are numerous and varied in unregulated markets and they tend to reduce considerably the return to the producer from the sale of his produce.[21] The farmer-seller of the study area has to arrange a *kachcha arhatiya* in the marketplace for the sale of his produce and thus he has to pay commission to *arhatiya*, *dalal*, weigher and labourers. Sometimes a seller has to submit to a deduction known as *garda*, *dalta* and *dhamada*. The objectionable feature about the market charges is that they are not only high but are also not clearly defined and specified. In most of the villages the storage facilities are poor because of indigenous methods of storing the grains. This area does not enjoy the warehouse and other storage facilities even in the adjoining big markets and *mandis*. Lack of information is another problem as villagers have practically no contact with the outside areas. They are not in touch with the trend of market prices and they mostly depend on hearsay reports received from the village traders. Moreover, this backward area does not enjoy the facility of such banks which could finance them at a low rate of interest against the security of goods.

Planning the Transportation and Marketing Facilities

In 1974-75 when the Patha area was recognised as a drought-prone area the State Government paid some attention towards transport development in the area and the following roads were proposed for construction (Fig. 16.4).

TABLE 16.4: Proposed Roads

	Name of road	Pla~es to be connected	Length (km)	Condition
1.	Dadri to Dewangna	Dadra, Madaiyan and Dewangana.	16	Primary work has been done
2.	Bahilpurwa to Ohan nala (Ohan Dam)	Bahilpurwa and Ohan Dam.	5	-do-
3.	Manikpur to Intwa-Dundaila	Manikpur, Baramuafi, Chh-eria Khurd, Kuhuniyan, Bambia and Intwa Dundaila.	31	-do-
4.	Bargarh to Madha	Bargarh and Madha	5.25	-do-
5.	Dadri to Markundi	Dadri and Markundi.		Proposed for construction in Sixth Plan
6.	Manikpur to Bargarh	Manikpur, Church Kesaruwa - and Bargarh		-do-

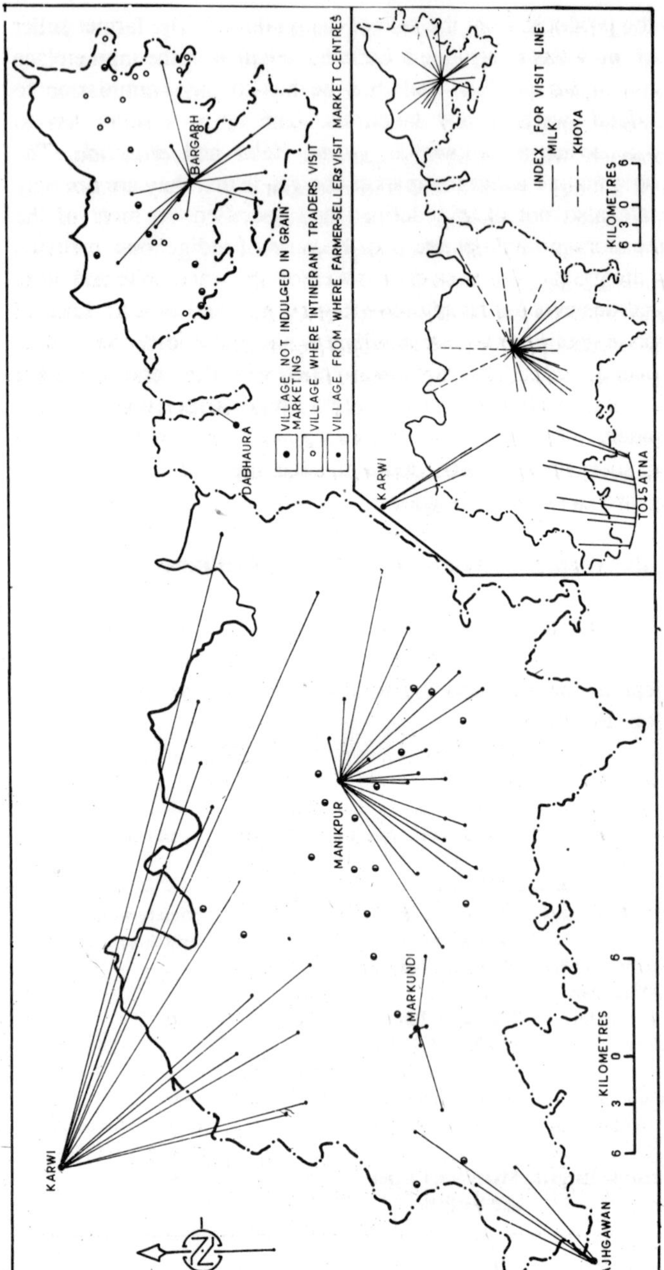

FIG 16.4: Patha Area : Grain Marketing and Visit Pattern

The present transport development programmes of the study area are oriented to project approach as transportation programmes in developing countries have been evaluated on a single project basis, and both inter-modal and intra-modal effects have been overlooked.[22] In project approach total system affects are not considered in the evaluation of a single project, and transportation plan is not related to the over-all economic plan. Thus in the absence of integration between the general economic plan and the transportation plan, the chief goals of transportation planning are neglected.

The basic premises which are to be kept in mind, while formulating a model for transport network, are to increase the locational utility of central places by providing accessibility to villages lying in the hinterland, to make central places serve as the focal points of the transport network, and to upgrade the different categories of roads through the progressive stages of spatial organisation.[23] While using this model in the present context the norms set by the National Planning Commission about the 'Minimum Need Programme' should not be overlooked. It was for the first time that in the fifth Five-Year Plan, rural roads were treated as a part of the 'minimum need programme' and in that plan the aim has been to link all the villages having the population of 1500 and above with all-weather roads and 50 per cent villages with a population between 1000 to 1500 to be linked by roads by 1985. For implementing this programme the State Government was to take all necessary steps to draw up realistic proposals taking an integrated view of the over-all requirements, technoeconomic aspects and social needs of rural economy.

Keeping in view the basic premises of the said model and the norms of 'minimum need programme' of transport, the development points of the study area must be interlinked to each other, in the first phase, in such a way that new roads would cut through the inaccessible patches. In the second phase the same principle may be extended taking the central villages as the basic points, and connect them to neighbouring villages which have the population between 500 and 1000. Moreover, those villages, which will attain a population of 1500 in 1991, should also be linked with all-weather roads. Village Rukma Buzurg, Hanuwa, Chandmara, Garhchapa and Unchadeeh, which have a population of more than 1500 at present must be linked by metalled roads to the nearby existing service centres or growth centres so that the villagers may not face difficulties in future in marketing their produce and in purchasing various inputs and other goods of periodic use. It is also suggested that the State Government must mobilise the resource so that the roads, which are still to be completed, must be ready by 1980 because the flow of input and output will definitely be increased more than double of the present position.

In order to implement the minimum need programme for rural transport development and minimise the cost of constructing new roads, some of the forest roads linking the service centres (Fig. 16.1), and central villages should be made open for private and government buses with the coordination of forest department. Moreover, some of the cart tracks linking with service centres or central villages should be converted into motorable road for the smooth movement of goods and passengers. The opening of forest roads and their conversion to metalled roads will make the flow of forest produce and minerals, such as bauxite, ochre and clay, easy. The large settlements and central places situated near railways should also be connected to railway stations by motorable roads so that a coordination between rail and road may be maintained. In order to develop inaccessible patches and make the interaction between villages and central places of higher orders more active than at present, the conversion, improvement and new construction of the following roads should be taken up by the State Government (Fig. 16.4).

TABLE 16.5: Roads Suggested for Conversion and Improvement

Sl. No.	Name of the road/cart track	No. of settlements benefited
(a)	*Conversion of forest road to metalled road*	
1.	**Manikpur to Katautha Mamania via Ranipur**	6
2.	**Manikpur to Rajauhan forest (for utilising minerals and forest produce)**	2
3.	**Ranipur to Rampur Kalyangarh**	4
(b)	*Conversion of cart track to motorable road*	
1.	**Markundi to Manikpur via Rampur Kalyangarh.**	7
2.	**Ranipur to Unchadeeh.**	4
3.	**Bargarh to Murka.**	2
4.	**Bargarh to Guniyan Kalan.**	2
(c)	*New short distance link roads*	
1.	**Kailaha to Dadri.**	—
2.	**Rukma Khurd to Bahilpurwa Rly. station.**	—
3.	**Lapaon to Dabhaura Rly. station (M.P.).**	—

During rainy season almost all villages, except a few situated along metalled roads, remain cut off from other parts of the district as well as neighbouring areas of other districts of U.P. and M.P. only on account of inferior quality of roads. Therefore, the quality of roads of different categories is to be improved for smooth and quick traffic flow. No doubt, after the completion of the proposed and suggested road network, this backward area will emerge as develooed area atleast in the sphere of transportation development.

Spatial organisation to the optimum level cannot be achieved unless a sound net of periodic markets as well as daily markets is woven in the study area. The periodic markets generally perform two parallel economic functions: one for local exchange and the other for provision of imported retail goods. For diversifying the human activities, facilitating the disposal of village produce and generating significant growth impulses to their surroundings, all the seven rural central places, i.e. Kehuniya, Saraiyian, Unchadeeh, Murka, Baramuafi, Rukma Khurd and Intwadundela (Figs. 16.1 and 16.4) must have periodic markets. For this very task the *adalat* panchayats must be persuaded to start periodic markets at these rural central places.

The weakest link in the rural development efforts in this backward area is organisational set up, particularly on marketing front. Infrastructural elements such as – transport, communication, irrigational facilities, electricity, and financial institutions are present in some measure but no marketing organisation exists in the area. According to Kuo-chin[24] 'a practicable organisation is as important as a good programme in countries like India, where the organisational base is very defective'. There is a great scope, therefore, to create an integrated organisation that will not only facilitate the marketing of agricultural products but also regulate the financial credit.[25]

In order to make marketing more efficient and to minimise the problems of storage and warehousing, establishment of market towns is the need of the day. Johnson has advocated the setting up of market towns for regularising rural economy.[26] Owen endorses the role of market towns for 'commercialisation of agriculture'. According to him "more market towns are necessary to provide farmers with a convenient place to bring their produce and to supply warehouse, processing plants, consumer goods and credit facilities."[27] Most of the defects and malpractices to the disadvantage of producer-seller can be removed by the exercise of proper control over markets and this could be done by the establishment of regulated markets either by local bodies or under State legislation. The setting up of regulatea market at Manikpur and Bargarh is prerequisite for the welfare of the farmer-sellers of the area. Manikpur must be developed as the main market town for improving the marketing facilities by establishing a 'mandi' and connecting it by all-weather motorable roads with the rural central places of Manikpur Patha area.

Cooperative marketing of agricultural and animal produce is necessary not only for establishing marketing conditions by means of orderly and regulated supply of commodities but also for improving the economic conditions of the producer by strengthening his bargaining.[28] Under different Five Year Plan periods cooperative marketing was encouraged in all possible ways but the marketing societies in India have not done well because of defective working. Inspite of the present condition of cooperative marketing primary marketing societies should be located at *mandi* centres and the area of cooperation should be the hinterland of the market concerned. As some of the cooperative societies are working well in the area the primary marketing society should be started to function at the earliest so that it may minimise the malpractices in the market centres and remove the intermediary system. Moreover, the establishment of multi-purpose farmers service society, which makes the farmers free from all botherations of obtaining credit, and of disposing of the product at remunerative price and thus allowing him to concentrate fully on increasing the farm productivity, must be expedited. This will also free the farmers from the clutches of itinerant traders and contractors. Also another important factor to be taken into consideration is the insulating of such cooperatives from the monopolistic influence of the big farmers and bureaucratic elements.

NOTES

1. Singh, J., "A Model for Rural Transport Network", in Singh, R.L. (Ed.). *Rural Settlements in Monsoon Asia*, Varanasi, 1972, p. 189.
2. Shivalingaiah, "The Role of Market Centres in Rural Development", in Masood, M.S. & Shivalingaiah, M. (Ed.) *Urban System and Rural Development, Part II*, Institute of Development Studies, Prasaranga, Mysore, 1972, p. 197.
3. Srivastava, R.C., & Ali, J. "Veterinary Hospital Facilities and their Planning in Drought-prone Areas: A Case Study of Banda District" , *Transactions Indian Council of Geographers*, Vol. VI, 1979, p. 1.
4. ——— "Identification of Basic Planning Units for Integrated Area Development: A Case Study of the Patha Area of Bundelkhand", *Transactions Indian Council of Geographers*, Vol. VIII, p.2.

5. Ulman, E.L., "Role of Transportation and the Basis of Interaction" in Thomas, N.L. Jr. (Ed.) *Man's Role in Changing the Face of the Earth*. University of Chicago Press, 1956, p. 875.
6. Singh, H.P., *Resource Appraisal and Planning in India*, New Delhi, 1979, p.113.
7. Kansky, K.J., *International and Interregional Comparative Studies*, Army Transportation Research Command, Fort Enstis, Verginia, 1962.
8. Taaffee, E.J., Morill, R.L., & Gould, P.R., "Transport Expansion in Underdeveloped Countries : A Comparative Analysis in Transportation Geography' in Hurst, M.E.E. (Ed.) *Transportation Geography: Comments & Readings*, 1974, p. 386.
9. David, N., "Some Aspects of Rail Community Network : London & Paris", *Geographical Review*, Vol. 49, No. 2, April, 1959, p. 153.
10. Singh, J., *Central Places and Spatial Organisation in a Backward Economy: Gorakhpur Region — A Study in Integrated Regional Development*, U.B.B.P., Gorakhpur. 1979, p. 61.
11. Verma, R.V., "Role of Periodic Markets in the Integrated Area Development: A Case Study of Safipur Tehsil of Unnao District", *Transactions Indian Council of Geographers*, Vol. VIII, 1980, p. 14.
12. Berry, B.J.L., *Geography of Market Centres and Retail Distribution* 1967, p. 106.
13. Pirenne, H. & Polyani, *Economic and Social History of Medieval Europe*, New York, Harcourt, 1936.
14. Mamoria, C.B., *Agricultural Problems of India*, 1977, p. 725.
15. Joshi, R.L. & Mamoria, C.B., *Principles and Practice of Marketing in India*, 1977, p. 636.
16. Bakshi, S.N. "Dairy Development in Drought-prone Area Programme" *Kurukshetra*, Vol. XXVII, No. 2, 1979, p. 20.
17. Singh, J., *Transport Geography of South Bihar*, Varanasi, N.G.S.I. 1964, p. 195.
18. Mukerjee, R.K., *Economic Problems of Modern India*, Vol. II, 1939, p. 295.
19. *National Planning Committee Report on Rural Marketing and Finance*, 1948, p. 42.
20. Munshi, C., *From Farmer to Consumer,* 1943, p. 34.
21. Mamoria, C.B., op. cit., p. 736.
22. Martin, B.V., & Warden, C.B., "Transportation Planning in Developing Countries" in Hurst, M.E.E. (Ed.) *Transportation Geography: Comments and Readings*, 1974, p. 408.
23. Singh, J., *op. cit.* p., 191.
24. Kuo-Chin Chao, *Agrarwirtschaftliche Entwicklung problems in hentingen,* Indien, Gudtav Fischer Verlarg. Stuttgart, 1964.
25. Singh, J. *op. cit.*, p. 117.
26. Johnson, E.A.J., *Market Towns and Spatial Development in India,* N.C.E., New Delhi, 1965.
27. Wilfred, O., "Distance and Development, Transport and Communication in India, *Transport Research Programmes,* The Brooking Institution, Washington. D.C., 1968.
28. Mamoria, C.B., *op. cit.*, p. 766.

PART V

GEOGRAPHY AND INTEGRATED RURAL DEVELOPMENT

17

SPATIAL ASPECTS OF RURAL DEVELOPMENT SCHEMES IN THE INDIAN DESERT

A.K. TEWARI

RURAL DEVELOPMENT is primarily concerned with stimulation of the total spatial system of an area through judicious use of natural resources for balanced regional growth. Recent studies have indicated a depletion in desert resources with a rising population which has spread the poverty in marginal areas and dissipated the material economic gains and the healthy process of planned development. This paper analyses the spatial aspects of rural development schemes in the desert and semi-desert areas of western Rajasthan, particularly the drought-prone area programme, desert development programme, desert national park and land development and irrigation schemes. The major source of information is the blend of socio-economic data published by the State agencies supported by the empirical evidences collected from the field. The analysis reveals that the desert region of India can be developed to the fullest extent not only by irrigation agriculture but also by organising and upgrading the non-crop economies and establishing new settlements with varied functions. Articulates system of communication and rural industrialisation will itself become an integral component of this process. The environmental constraints, regional and sub-regional disparities and local qualms will disappear under the new horizon of overall development.

Desert: The Problem Area

Apart from the institutional negligence and late commencement of modern planning, some of the basic problems of the desert areas are inherent to its environmental attributes. Arid environment is wedded to scarcity, especially of water and life-forms depending on water. Evidently this region is negative to human occupance and expensive to develop in the modern sense. Inadequacy of resources can be counter-balanced by special rural development planning which is in tune with the arid ecosystem and the aspirations of the people inhabiting this area. Traditionally, these are empty spaces with few people living under nomadic agropastoral economies. The settlements consist of a few oases near some water bodies which served as entrepots along the ancient carvan routes. The scattering and dispersal of these route-side trade points was proportional to the density of population of these areas and the frequency of carvan routes.

But the advancing technology and population spillovers from humid areas have progressively changed the original mosaic of occupance. Irrigation agriculture became a viable instrument of transformation of these areas, first in the riverine tracts of assured water supply, and later, in the hard core deserts. Artificial reallocation of surface water was made possible by inter-basin transfer of water from the Punjab plains to the desert region of northwest India. The three major irrigation schemes, namely Ganga Canal, Bhakra Canal and Rajasthan Canal, have had a varying degree of impact in the development of this region. Nature too remained at work in transporting groundwater from distant catchment areas to the aquifers buried deep in the desert sand. This, no doubt, produced a major modification to the entire desert ecosystem. By flow canals the dried up basins of Gaggar and Saraswati were saturated to the extent that waterlogging became a chronic problem in the midst of Thar desert.

Recent studies in the desert development prospects of India have indicated that much of the problem is man-made (Fifth Five Year Plan: Rajasthan-1974-79). The unorganised proliferation of agriculture in water thirsty areas is a major cause of desertification. These are the regions of meagre resources. The margin between human success and failure is so narrow that a minute deviation in precipitation or misuse of land may trigger off disasterous droughts and famines not only in the desert area itself but also in other areas adjoining it. For a long period of Indian history the people of this region lived under the great stress of drought and uncertainty. The delicate balance of arid ecosystem can be maintained by physical responses of drought resistant biomass and human activities using minimum possible water.

Water is pre-eminently an economic reserve which is to be wisely spent on alternative uses. Traditional agriculture consumes and evaporates large amounts of soil moisture — a scarce commodity, the shortage of which makes these areas into deserts. The so-called rural development miracle in the irrigated areas of the Ganga Canal colony in the northern fringe of the desert is really over-estimated because it was coupled with a package of input complexes — irrigation, improved seeds, chemical fertilisers and incoming vigorous stock of farmers from neighbouring areas. The effect was concentrated only in selected areas with selected crops. Moreover, the fruits of agricultural gains ultimately accumulated to the group of people who supplied technological innovation and improved inputs. Thus it could not make much of a dent in the aggregate level of development of rural areas. The structural change of an economically active population is necessary to boost the productivity of both land and labour. It can be achieved more vertically in urban type of activities linked with agropastoral raw materials. Recognising the needs of this region, the State and the Central Government have launched four rural development schemes which add a new dimension to the spatial transformation of the Indian desert:

1. The Drought Prone Area Programme
2. The Desert Development Programme
3. The Desert National Park, and
4. Land Development and Irrigation Schemes.

Drought Prone Area Programme

Since the beginning of modern planning, the western arid districts of Rajasthan have been treated as special problem areas. Considering the impact of irrigation to be limited in areal extent, a new scheme of development for chronically drought-prone areas was launched under the central plan scheme known as Drought Prone Area Programme.[1] In Rajasthan nine districts in desert regions (Jodhpur, Nagaur, Pali, Jalore, Barmer, Jaisalmer, Bikaner, Churu and Jhunjhunu) and four districts in the subhumid region (Banswara, Dungarpur, Udaipur, and Ajmer) were selected for assistance under DPAP. These districts and Tahsils constitute more than 56 per cent of the area and about 13 per cent of the population of Rajasthan. On systematic basis this programme was started in 1974-75 and allocations have been stepped up year after year as the work has progressed.

Special projects of small irrigation works, rural resource development, (mineral, pastures and livestock development) and diversification of village economy by reviewing traditional crafts, cottage industries and ancillary activities were undertaken as a part of integrated rural development. During the Fifth Five Year Plan period about Rs. 450 million were spent on DPAP in Rajasthan by the Central Government alone. There is every possibility of upward expenditure during the current plan period. The major thrust of development efforts is to create rural infrastructure of a permanent nature with a dynamic potential for multiplying production, generating employment, accelerating economic growth, eradicating sub-regional imbalances and restoring the ecological equilibrium of the entire region. This approach has an inherent fervour of growth with social justice, i.e. the fruits of programmes must reach the weaker sections of society such as landless labourers, marginal farmers and small-scale craftsmen.[2]

Desert Development Programme

During the Fifth Plan period the State Government sponsored a new scheme of Desert Development Programme (DDP) for eight districts of western Rajasthan. On the basis of potential for economic development these districts have been subdivided into two development units : (i) extremely arid (*Marusthali*): Barmer, Jaisalmer and Bikaner, and (ii) the arid (*Bagar*): Churu, Nagaur, Jodhpur, Pali, and Jalor. In the extremely arid areas planning schemes are exclusively for livestock development, preservation of natural pastures and enhancement of cattle feed and fodder supply. In the arid areas dry farming and organisation of rural marketing of dairy products, wool and mutton are the major objectives. The Rajasthan Government has prepared a core programme to increase 30 per cent of the income of rural population and diverting 20 per cent of the rural work force to non-crop primary activities. For this purpose more than Rs 1,000 million (including DPAP) has been spent in the State.

It is the greatest fallacy of former planning that more emphasis was given to food grain production without taking into account the economic return of agricultural output in an arid environment. During a good year of rain the production of food grain is more or less higher than the local demand, but in the years of monsoon failure the loss of agricultural production is to the extent of 90 per cent. On the other hand, the fall of production derived from animal husbandry is hardly 30 per cent during the years of severe drought. The local inhabitants also perceive agriculture as the most risky operation for desert areas.[3] The strategy of rural development ought to consider land-use capabilities and economic return from crop or non-crop rural operations. A study made in Jodhpur reveals that one acre of land devoted to food grain crop *(Bajra)* yields on an average a net profit of Rs. 50 a year, whereas the same land devoted to stock raising may yield Rs. 100 a year with lesser degree of susceptibility to drought and depletion of soil. The objective must be to increase the income of the farmers by a combination of dry farming and animal husbandry.

The DDP has made it a defined strategy to give a lower priority to agriculture except in areas of assured irrigation. The native *Paliwal* settlers of Jaisalmer have for centuries preserved the soil-water balance by indigenous water harvesting technique known as *Khadin*. Under this technique surface water is allowed to percolate into ground by embanking the upland catchments of gravel outcrops and sandhills. Their entire farming is dry-farming. They are one of the skilled raisers of camels in India. They are least susceptible to famines during years of drought because they have the wisdom to live in the deserts. The historical records are the testimony of the fact that *Paliwals* have lived in the desert areas whereas many other farmers have migrated to the better watered plains of east Rajasthan adding to the problem of over exploitation of soil in the semiarid desert fringe. The Desert Development Programme is one step forward to the traditional practices of developing these areas without importing water for irrigation.

The two major schemes of DDP — pasture development and dairy development — have already won enough credibility and understanding. Being in tune with the people's aspiration they give a real source of satisfaction to the local people. Villagers have no fear of any socio-psychological frustration which usually follows major irrigation works whose benefits are reaped more by the new migrants. By a simple programme of milk collection at farmers' door, the per capita income of 20 per cent villagers in Jodhpur and Bikaner Districts has increased on an average by about Rs. 200 per year. Many families which have never produced milk are new coming up as good suppliers of milk to the collection booths and chilling plants set up at smaller cities like Merta, Pali, Pokharan, Balotra, Lunkaransar, Sardarshahar, and Jhunjhunu. The planned spatial organisation of this activity may lay the foundation of several dairy complexes at regional centres like Jodhpur and Bikaner.

The Desert National Park

The National Commission of Agriculture has added a new dimension to the rural development and preservation of ecological balance of desert areas by recommending a Desert National Park. Apart from a purely tourist attraction the park will help understanding the life cycle and critical ecosystem of the Indian desert and will provide a natural habitat for endangered species of arid wild life (Great Indian bustard, Chinkara, spiny tailed lizards, desert cats, foxes and wolves). The desert wildlife thrives on tender shoots of grass-bushes, berries, and small animals like grass hoppers, beetles, centipedes, lizards, snakes and locusts.[4]

The State wild life department has taken over a plan of setting up a desert sanctuary in an area of 3000 sq.km in the districts of Jaisalmer and Barmer. The Union Ministry of Rural Reconstruction has allocated Rs. 25,000,000 for this park. Compared to the other national parks of the world such as Canadian National Park (80,000 sq.km), Tsave National Park, Kenya (8,000 sq.km) and Serengheti National Park of East Africa (5,000 sq.km) the proposed ecounit is pretty small. Yet a modest beginning has been made. The native Bishnois of Rajasthan have helped in preserving desert deer because of the religious sanctity given to this animal. The Bishnois believe that deer are their ancesters who have taken rebirth in that form. They erect water pans on the outskirts of their villages where every Bishnoi has to offer water for deer at sunset. Many of the village level quarrels among the Bishnois and the Rajputs (ruling caste) stems from the deer hunting by the latter. The desert national park will be free from any human intervention — hunting or farming. All the natural pools and *thals* (basins) will be developed into winterhomes of bustards and imperial grouse which land here in millions during November every year. On the shores of these water bodies colourful flowers of *capparis decidua* and trees of *rohirrka* bloom. The landing of several migratory birds on the shores of small *thals* is a phenomenon unparalleled from the bird-watcher's point of view.[5]

The scheme has a three-stage operational programme involving a total expenditure of about Rs. 35 million over a period of five years. Firstly, the area of the national park will be classified as a core zone, free from any intervention of human activities. It will be surrounded by a peripheral belt of controlled grazing and restricted farming. It will be supplemented by new plantations of thorny bushes and supporting flora which will provide food as well as camouflaging space for the threatened desert species. Secondly, a research centre will be developed within the park to conduct special studies of the desert flora and fauna mainly for their ecological problems. A small study centre has already been established to monitor the behaviour and to identify the bee species of desert terrain capable of storing large honeycombs for a longer period. Finally, an intricate network of tourist observation posts, camps, ***machans***, and tracks will be developed in the entire region connecting with the larger settlements like Jodhpur and Bikaner.

Under this scheme the Government of Rajasthan has added a new attraction. At Akal (some 17 km from Jaisalmer) a fossil wood park has been planned where all the fossil wood logs of Lower Jurassic period will be preserved as a national museum. This region is famous for fossil woods of the geological past. The Geological Survey of India has so far counted more than 60 pieces of fossil logs buried under the desert sand. The biggest of such finds measures 7.5 metres long and 1.5 metres wide. It is believed that these fossils have been deposited in this area by wild rivers millions of years ago during the Lower Jurassic period. During the Permocarboniferous period this region was covered by dense forests of non-flowering type, like pines. The Pleistocene desertic conditions and recent ground water circulation in the fossil rich area has manifested an architectural design of fossil wood which resemble the superb carvings of Jaisalmer.

Land Development and Irrigation Schemes

Since the desert regions receive rainfall much below the water requirement for most of the agricultural plants, irrigation appears to be a necessity for farm operation. During the princely period inter-state water transfer was not possible due to political fragmentation and lack of financial coordination. The irrigation work was then confined to impounding the local run-offs into tanks and reservoirs. In the northern part of the arid region the Ghaggar bed has now been fully utilised by the canal network whose only source of water is the Sutlej river of Punjab. This planned development of hydraulic resources commenced in 1927 by the construction of the Ganga Canal and later it was augmented by huge multipurpose schemes of the Bhakhra project and the Rajasthan Canal project. The Ganga Canal takes its water from the Hussainabad-Firozpur Head (now linked with Hari Ke Barrage), and the Rajasthan Canal from the Hari Ke Barrage near the point of confluence of River Beas. The flow water has further been augmented by linking the entire system with River Ravi near Pong Storage. It is really the most stupendous and ambitious area development scheme of our country. The lessons from the Ganga Canal project has made it necessary to plan the Rajasthan Canal Project as a regional development model for total transformation of the rural landscape in an area of sandy waste. The spectacular achievement in canal irrigation in the desert districts made it possible to increase the irrigated area from a meagre 600,000 acres in 1949 to nearly 1,600,000 acres in 1980.[6]

The agropastoral impact of land development and canal irrigation is already on the canvas of desert areas with more expedient crops and more complete use of land during the whole vegetal period. In addition to purely agricultural changes, it has called for development of roads, cities and villages. The level of agricultural development is so high in the Ganga Canal Colony that its population increased from 29,000 before the construction of the Canal to about 1,000,000 in 1981 (estimated). Similarly, in Rajasthan Canal commond area about 3,000,000 people will be settled into twelve thousand new villages, two hundred service villages and market centres. In essence the Rajasthan Canal project calls for transforming a region marked by pastoral nomadism and an impoverished rural economy to a major food producing region. The effect of these changes will be reflected in the landuse pattern within a decade. The villages will become larger and more stable, communication between them will increase, and their isolated character will give way to participation in the national economy. Out of the projected capacity of irrigating 1,300,000 hectares of land, the Rajasthan Canal is already irrigating about 300,000 hectare of land in Bikaner district (mainly Suratgarh State Farm). Some of the crops such as gram, cotton, wheat, sugarcane and mustard unknown in this area till recently, are being more successfully raised.

Together with economic development, many regional problems have also crept in as a result of the introduction of canal irrigation and land development on such a large scale. The demands of a developing society are often too great to be tackled only by agriculture. The influx of large number of refugees from West Punjab, especially following political unrests, has created an extraordinary situation in the Ganga Canal Colony. The entire spending on expensive scheme have been absorbed by the incoming migrants leaving almost no scope for the small farmers depending on dry farming. Over a period of few decades the local inhabitants have become casual labourers on the big farms owned by more energetic and versatile newcomers. Although yields have shown spectacular increase yet its efficacy as a means of disbursing social equity has nowhere been achieved. This has lead to a great degree of economic disparity within the canal colonies. The increase in production lowered the prices, frustrated the marginal farmers and outrated them in competition. Since social education was not included in the scheme, the economic affluence lead to waylayed spending and increase in crime rate. Thus it is clear that the planners of major irrigation schemes must consider socio-cultural parameters along with physical planning.

In the Rajasthan Canal Project area such problems are not apparent on the surface because greater part of the area is unpopulated. Yet social planning and sensitive compromises with the cattle and livestock raiser is necessary. As seen in the Ganga Canal Colony, the life style and social behaviour of the local people of the Indian desert will also change as a result of improved economy. The nomads will become sedentary, the subsistence farmers will experience new cycles of farming, new crops and a new dependence on market economy. The nuclear settlements will become larger with increased inter-village interaction. The small towns of western Rajasthan will grow demanding more tertiary facilities depending on the outside resources upon which the local people would have little control. Bureaucratic uncertainties and lapses of canal officials will create uneasy social relations. With the introduction of cash crops, the national and international economic situations will shape the rural economy which has to depend on outside agencies for modern inputs and sale of surpluses. The interbasin transfer of water from the Himalayas to an area of inland drainage will bring many ecological problems. To avoid the process of desiccation an outlet will be required from the fag end of Rajasthan Canal to the Arabian sea. To preserve the economic autonomy of the existing villages and existing cities, secondary and tertiary activities will be required on a larger scale. All such changes can be comprehended in an atmosphere of perfect political peace, international cooperation and judicious management of both rural and nonrural resources.

NOTES

1. Agro-economic Research Centre, *Levels of Agricultural Development in Tehsils of Rajasthan*, Research Study No. 42 (Mimeographed), Vallabh Vidya Nagar, 1975, pp 10-16.
2. Govt. of India, *Integrated Agricultural Development in Drought Prone Areas, Report of Task Force Committee on Integrated Rural Development*, Planning Commission, June 1973, p 42.
3. Tewari, A.K., "Agricultural Landuse in the Arid tract of Rajasthan," *Landuse in Developing Countries, Proc. of the 21st IGU Symposium on Landuse in Developing Countries*, Aligarh, 1972, pp 24-31; Perception of Drought Hazard Among the Farmers of Indian Desert; *Perspectives in Agricultural Geography : Human Dimensions in Agriculture*, Vol V, Concept Publishing Co., New Delhi, 1981, pp 223-40.
4. Govt. of India, *Rainfall and Cropping Pattern — Rajasthan* Vol. XIII, National Commission on Agriculture, Ministry of Agriculture and Irrigation, New Delhi, 1976, pp 20-25.
5. Govt. of Rajasthan, *Draft Fifth Five Year Plan 1974-79, Rajasthan*, Planning Department, Jaipur, 1973, pp 61-65.
6. Shastri, G.S., "Rajasthan Canal Project; A Dream of the Century," *Bhagirath*, The irrigation and power quarterly, Vol. XXIV, No. 3, Special Issue on Combatting Desertification, July 1977, Central Water Commission, New Delhi, 1977, pp 11-17.

18

LOCATION AND ALLOCATION OF CENTRAL PLACES
A Study in an Integrated Area Development of Bijapur District (Karnataka)

F. V. YAVAGAL

INTEGRATED AREA Development not only aims at integrating social and economic activities which influence the life of the people but also aims at optimal location of investments so as to reduce the gap between developed and developing areas (Misra, R.P.: 1971). Economic growth is not accomplished merely by providing material input. It is also necessary to ensure that the investments should be made at the right places in the right manner according to the need, and capabilities to sustain and serve the concerned areas. In the space, there are few favoured points around which services and functions tend to cluster. Here an attempt is made not only to identify such places along with their hierarchical orders and functional levels but also to study their spatial patterns to bridge the gap between functional and spatial disparities in Bijapur District of Karnataka State.

Study Area

Bijapur District (15° 20′ to 17° 28′ N latitudes and 75° to 76° 30′ E longitudes) has an area 17,050 sq. km and a population about 2 million (in 1971). It comprises 1285 settlements out of which 17 are urban. The district has two contrasting areas – Northern and Southern. The southern part of the district (south of the Krishna) is well drained by the Krishna and her tributaries – the Ghataprabha and the Malaprabha. It is densely populated area with fertile soils and is developing very rapidly due to irrigation projects on all the three rivers. Just the opposite is true for the north (Fig. 18.1).

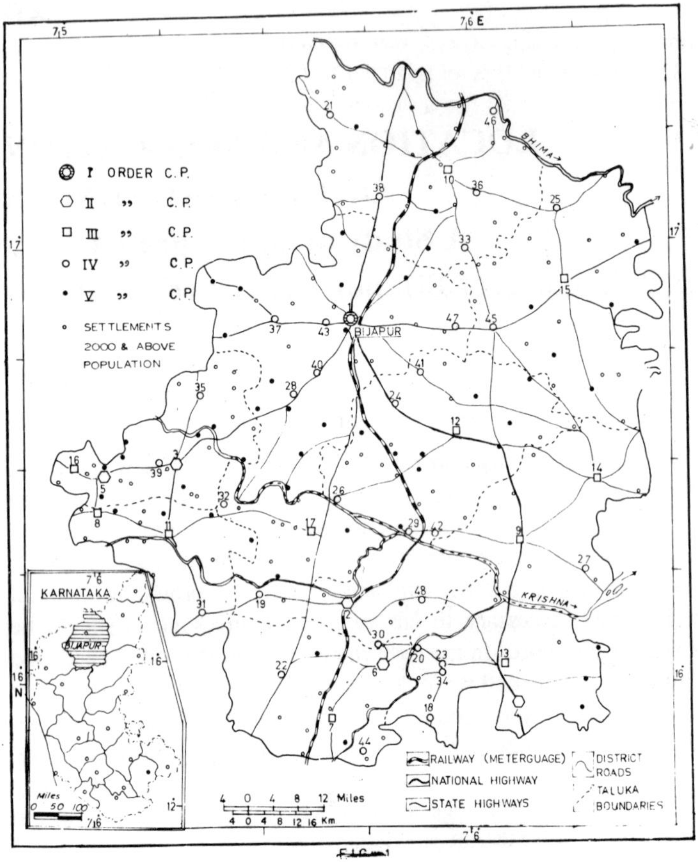

FIG 18.1: Bijapur District : Spatial Pattern of Central Places

Identification of Central Places

In the present study 8 basic determinants along with their 22 internal hierarchical levels and 4 non-basic determinants (Table 18.1) were considered as indicators to identify the central places. The hierarchical order of an individual function or service within the basic/non-basic determinant was fixed on the basis of its relative importance and specialisation. Then the sub-functions/services were arranged in an order and weighed according to the geometric progression. These weightages were multiplied by their strength – the number of functions/services existing in the settlements under consideration. Then summed-up centrality scores were derived for all the determinants for the 328 settlements. It is an accepted fact that all functions and services are not available in all the settlements of all sizes. So 328 settlements having population of 1500 and above (the average size of the settlement is 1545) were examined in detail. The settlements which scored more than 25 centrality scores were identified as central places. There were 100 central places in the entire district (Appendix I).

Hierarchy of Central Places

Five-tier Hierarchical orders of the central places were observed in the district (Fig. 18.1). There was only one First order central place – namely Bijapur city, with the summed-up centrality score of 1844. There were 5 Second order central places with scores ranging from 232 to 64,811 Third order centres with the centrality range of 120 to 198, 31 Fourth order centres ranging between 50 and 100 score values, and 52 Fifth order central places of 25 to 48 centrality scores (Table 18.2).

Hierarchy of Functions/Services

The notion that higher order centres have higher level functions was also varified in the study area. Five levels of functions/services were closely associated with the five hierarchical class system. Full range functions (F) exist only in that and above that order of centres, and below that level either they are partial (P) or some (S) or some times none (Tables 18.3 and 18.4).

Spatial Pattern of Central Places

One of the important aspects of the central places is their relative locations in space. They should be distributed hexagonally (uniformly) to serve the area properly. To study this distributional aspect nearest neighbour analysis, originally developed by two botanists, Clarke and Evanue (1954) has been applied to know whether there exists random

TABLE 18.1: Measures of Functions, Sub-functions and Their Weightages

	Sl.No.	Sub-Functions/Services	Weightages
Basic Functions/Services			
I Education	1.	Primary Schools	2
	2.	High Schools	4
	3.	Composite and Jr. Colleges	8
	4.	Sr. Colleges and Technical Institutions	16
II Health	5.	Medical practitioners	2
	6.	SMP/PHC/FP	4
	7.	TDB/Municipal/Private nursing homes	8
	8.	District Hospital/specialised institutions	16
III Agriculture and Allied Activities	9.	Small scale and other societies	2
	10.	Large scale and Farmers co-op. societies.	4
IV Veterinary Facilities	11.	Veterinary Aid Centres	2
	12.	Veterinary Dispensary	4
	13.	Veterinary Hospital	8
V Marketing	14.	Weekly Markets	2
	15.	Regulated Markets	4
VI Communication	16.	Post Offices	2
	17.	PTO/Telephone offices	4
VII Commerce and Finance	18.	Grameen Banks/Pay offices	2
	19.	Bank Branches/Co-op. Banks	4
	20.	Head Offices/DCC Banks	8
VIII Recreational Facilities	21.	Public Libraries	2
	22.	Cinema Talkies	4
Non-Basic Functions/Services			
IX Other Amenities	23.	Electricity	2
	24.	Drinking water facility	2
X Road Accessibility	25.	Accessible to District roads	2
	26.	State Highway	4
	27.	National Highway/Rly. Station	8
XI Road Nodality	28.	Single district crossroad	2
	29.	Double crossroads with State highway	4
	30.	Double crossroads with SH/NH/Rly.	8
XII Location Quotients of Tertiary Activities	31.	Less than the LQ	2
	32.	+1 LQ	4
	33.	+2 LQ and above	8

TABLE 18.2: Size-wise Distribution of the Central Places in Bijapur District

Zone	I	II	III	IV	V	Total
South of Krishna	00	5	6	12	18	41
North of Krishna	1	00	5	19	34	59
District	1	5	11	31	52	100

TABLE 18.3: Orders of Central Places and Associated Groupings of Central Functions on the Basis of Their Range in Bijapur District

Levels of Central Functions	Orders of Central Places			
	I	II	III	IV
I	F	S	O	O
II	F	F	P	S
III	F	F	F	S
IV	F	F	F	F

F = Full range, P = Partial range and S = Some range.
If a function occurs once or more than once per centre per function then it is full range, if it is between half to one it is partial range and if it is less than half it is some range.

TABLE 18.4: Functional Levels in the Bijapur District

V Level Functions related to V Order Centres

1. Lower Primary School
2. Post Office
3. Electricity
4. Drinking water facility
5. Medical Practitioner
6. Pakka road

IV Level Functions related to IV Order Centres

7. PTO and Phone
8. Medical Practitioner within door facility
9. Weekly market
10. Small scale co-op. society.
11. Higher Primary school.
12. District roads.

III Level Functions related to III Order Centres

13. Grameen Banks
14. High Schools
15. SMT/PSC/FP
16. Veterinary aid centres
17. Regulated markets
18. Large scale and Formers co-op. Societies.
19. Public Library
20. Town Panchayat.

II Level Functions related to II Order Centres

21. Bank Branches
22. Cinema Theatres
23. TDB/Municipal/Private nursing home
24. Veterinary dispensaries
25. Jr. & Sr. Colleges
26. Municipalities

I Level Functions related to I Order Centres

27. District Health Office and Specialised institutions
28. DCC Banks
29. Veterinary Hospitals
30. Administrative headquarters.

pattern or clustering or uniform pattern. "Nearest neighbour technique is not simply a statistical 'toy' to be played around with, but a genuine tool to improve the analysis that geographers make of the settlement distribution pattern (Ambrose: 1922 and 1972). This technique involves simply a comparison between two mean magnitudes: the observed mean distance between the nearest neighbours ($\bar{r}O$) and the expected mean distance of the same density of points if they were distributed randomly ($\bar{r}E$). This $\bar{r}E$ can be obtained on the basis of the formula:

$$\bar{r}E = 0.5\sqrt{\frac{A}{N}}$$

where A is area and N is number of central places. Nearest neighbour index is the ratio of these two ($Rn=\bar{r}O/\bar{o}E$). A nearest neighbour index of one or near to one would indicate that the observed pattern was like a random pattern. Deviation above one reaching a maximum of 2.15 shows regular pattern and deviation below one reaching a minimum of *O* shows the clustering pattern of central places. The method was applied separately to the two distinct areas of the district (south of the Krishna and north of the Krishna, Fig. 18.1) which differ in all respects and needed separate attention. The nearest neighbour index of 1.37 for 59 central places distributed in the northern part of the district showed a more regular dispersal of the places. Regular component is more significant than the random component in this area. But in the south, the distributional pattern of 49 central places was like a random with *Rn* value of 1.08. Here random element is more significant than the others. Various natural, socio-economic and cultural factors might have acted in various directions with varied magnitudes pulling some along the rivers and roads and pushing some to the corners, finally resulting in a random-like pattern in the southern part which is very well drained by the Krishna, the Ghataprabha and the Malaprabha rivers. In the north, monotonous 'plateau' topography and drought-proneness, weak transportation links and lack of raw materials might be the responsible factors for uniform distribution of the central places.

Functional Gaps

As the identification of the central places and their hierarchical order was done on the basis of the existence of 'Full Range' functions/services of that and below that level, there are seen only a few gaps in the higher order central places. Functional disparity has been observed in terms of their magnitude in higher order places, while in some of the lower order central places they are either partial or some, or some times none. The southern area needs two more specialised medical institutions, one at Jamakhandi and the other at Ilkal; bank headquarter at Jamakhandi, and two more veterinary hospitals at Guledgud and Ilkal so as to serve the people and the bovine population properly. Fourth order centres need 5 telephone and P&T Offices, 7 Medical Services, 9 weekly markets and a small scale co-operative society at Savalagi. Three centres – Savalagi, Tikota and Hunnur – are to be connected with electricity to call themselves 'Fullfledged' central places of that level. More than 1/3rd of the fifth order central places need many services and amenities of that level.

Spatial Disparities

There is a wide range of difference between the southern and northern parts of the district. All the five second order and about half of the third order central places are located in the south which covers about one third of the area. None of the centres in the vast area of 10,900 sq. km of the north qualify themselves as second order centres. There are wide areal gaps which need to be served with varied levelled central places and related functions.

Spatio-functional Gaps

There was a greater 'felt-need' to identify the spatio-functional gaps so as to bridge the gap between the two for overall development of the district. Here an attempt is made to identify the spatio-functional gaps on the basis of the magnitudes (centrality score values) of the central places. The average area served by each central place of each class system (order) has been calculated on the basis of the formula:

$$\overline{A}_i = \frac{Ta}{Ts} \times \overline{S}_i$$

where $\overline{A}_i$ – the average area served by *ith* order *C.P.*
T_a – total area of the district
T_s – total centrality scores
$\overline{S}_i$ – average score values of *ith* order central places.

TABLE 18.5: Average Size of the Central Places (Score Values — Classwise)

Zone	I	II	III	IV	V
S.K.	00	352	161	97	36
N.K.	1844	00	166	65	33
District	1844	352	163	69	34

TABLE 18.6: Average Area Served by Central Places (based on score values) (Classwise sq. km)

Zone	I	II	III	IV	V
S.K.	00	505	232	112	51
N.K.	4004	00	360	141	71
District	3380	644	300	127	62

Then the circles were drawn on each central place of that order and above to show that that area is not only served by a central place of that and above order but also by that particular (levelled) functions/services. Thus the areas served by each central place of each order along with the related functional levels were mapped (Figs. 18.2 to 18.5). The unserved areas are the spatio-functional gaps which need proper attention. Proposals are made to reduce these areal disparities exiting in the district (see above referred maps).

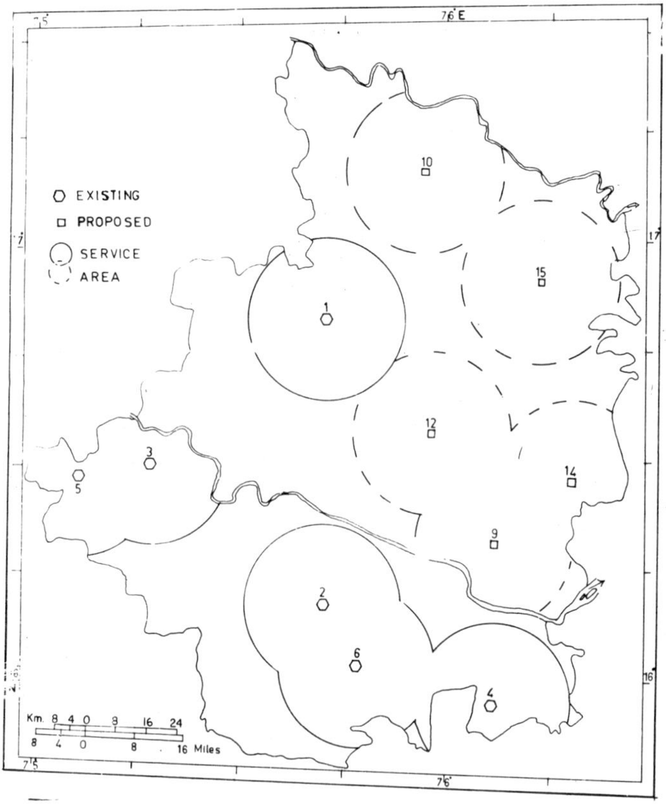

FIG 18.2: **Bijapur District : Location and Allocation of Second Order Central Places**

Central places not only serve the people of their own but also serve an additional population residing in the surrounding areas. While alloting the functions/services and planning for an integrated area development one should also consider the surplus population which it serves. The surplus population was calculated on the basis of the formula –

$$S_i = \frac{P_t - P_c}{T_s} \times \overline{S}_i$$

where S_i – surplus population served by each of the *ith* order C.Ps.

P_t – total population of the area

P_c – total population of all C.Ps.

T_s – total centrality score values

$\overline{S}_i$ – mean magnitude of the *ith* order C.P. (average centrality score values).

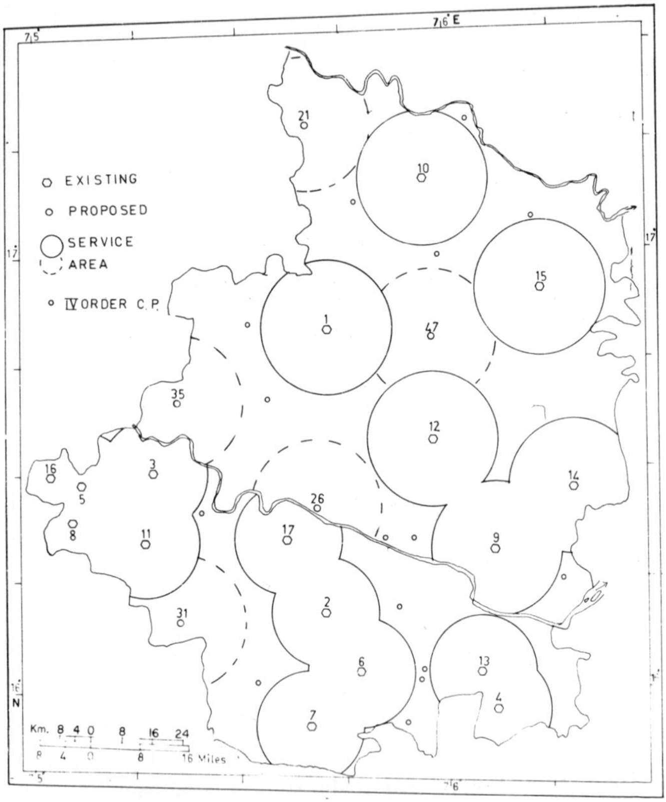

FIG 18.3: Bijapur District : Location and Allocation of Third Order Central Places

So Fifth order centres serve an additional average population of 4,174 along with their own average population of 3,737 (Tables 18.7 and 18.8) in the district. The existing services and amenities are not sufficient to serve an average population of about 8000. Therefore, the fourth level functions should be extended to Fifth order centres. Likewise, Fourth, Third and Second order central places serve an average population of 15,000, 32,000 and 80,000 respectively. The existing function/services are incapable of serving such a large population. Therefore, there is a greater need to extend the third level functions to fourth order centres, second level functions to third order central places (it is more true for the northern part of the district which lacks in second order central places). There is also the need for the development of the 'fullfledged' regional headquarter in the south.

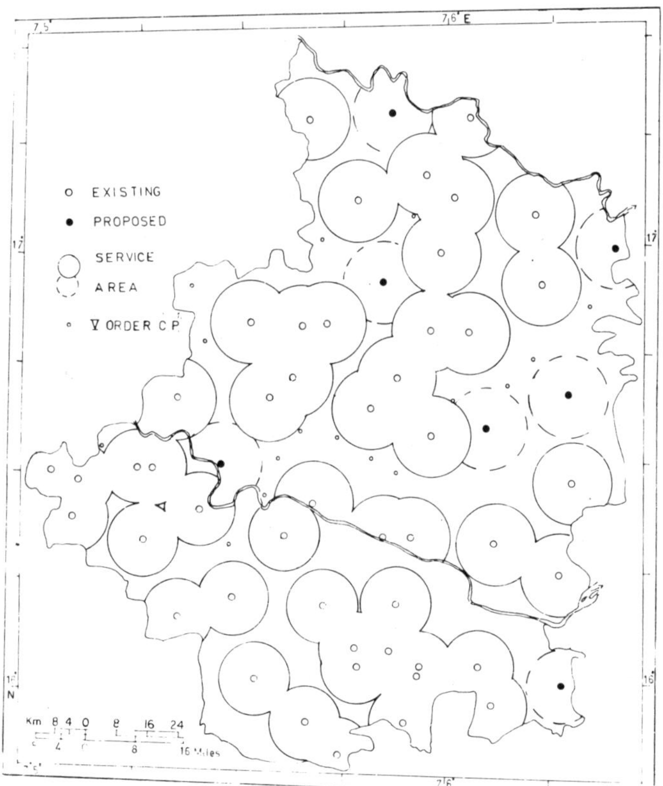

FIG 18.4: Bijapur District : Location and Allocation of Fourth Order Central Places

TABLE 18.7: Average Size of the Central Places (Population — Classwise)

Zone	I	II	III	IV	V
S.K.	—	35123	12591	6584	3550
N.K.	103931	00	12577	6171	3836
District	103931	35123	12535	6331	3737

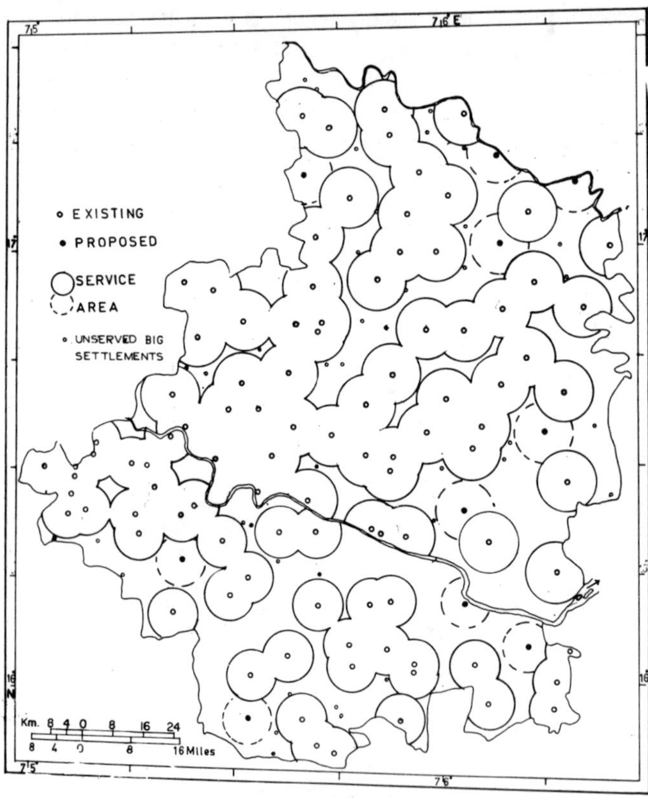

FIG 18.5: Bijapur District : Location and Allocation of Fifth Order Central Places

Recommendations

There is a greater need for an integrated area development plan for the Bijapur district. Based on the foregoing analysis the following recommendations may be made with a view to reducing the functional and spatial disparities prevailing in the study area:

1. Bagalkot should be upgraded as 'fullfledged' regional headquarter.
2. Second level functions/services should be extended to Indi, Sindgi, Basavana Bagewadi, Muddebihal and Talikot.
3. Chadachan (Indi), Savalagi (Jamakhandi), Kolhar (Basavana Bagewadi), Shivanagi (Bijapur) and Lokapur (Mudhol) should be provided with third level functions.
4. The following Fifth order central places should be upgraded and provided with fourth level functions:

   (i) Nagathan (Bijapur)
   (ii) Halasangi (Indi)
   (iii) Moratagi (Sindgi)
   (iv) Katakeri (Sindgi)
   (v) Salwadgi (B. Bagewadi)
   (vi) Bidari (Jamakhandi)
   (vii) Nandawadagi (Hungund)

TABLE 18.8: Average Surplus Population Served by Each Central Place (Classwise)

Zone	I	II	III	IV	V
S.K.	00	40434	18550	11120	4023
N.K.	251337	00	22571	8919	4484
District	227734	43423	20168	8571	4174

5. The following rural settlements should be developed as Fifth order central places:

Name of the Settlements	Population	Location code No.
(a) Neelgund (Badami)	2632	58
(b) Sasanur (B. Bagewadi)	2373	36
(c) Karadi (Hungund)	2620	40
(d) Sangam (Hungund)	2504	32
(e) Jagajeevani (Indi)	3369	118
(f) Khedgi (Indi)	2259	11
(g) Basarkod (Muddebihal)	7899	62
(h) Ingalagi (Mudhol)	2149	71
(i) Balaganur (Sindagi)	3780	89
(j) Devangao (Sindagi)	3734	58

APPENDIX I

Five Tier Hierarchical Order of Settlements in Bijapur District.

Rank	Name of Settlements	Score values	Population
1	2	3	4
First Order Central Places			
1.	Bijapur	1844	103931
Second Order Central Places			
2.	Bagalkot	648	51746
3.	Jamakhandi	334	29981
4.	Ilkal	278	30014
5.	Rabkavi-Banhatti	266	37507
6.	Guledgud	232	26365
Third Order Central Places			
7.	Badami	198	11651
8.	Mahalingpur	192	12635
9.	Muddobihal	188	10422
10.	Indi	180	13745
11.	Mudhol	172	15612
12.	Basavana Bagawadi	162	14002
13.	Hungund	160	12683
14.	Talikot	154	14537
15.	Sindgi	144	10178
16.	Terdal	126	15337
17.	Bilgi	120	7629
Fourth Order Central Places			
18.	Gudur	100	7380
19.	Kaladagi	100	7264
20.	Kamatagi	98	10144
21	Chadehan	98	6940
22.	Kerur	96	10820
23.	Amingad	90	7855
24.	Managoli	80	8383
25.	Almel	80	7629

Contd.

APPENDIX *Contd.*

1	2	3	4
26.	Kolhar	80	6767
27.	Nalatwad	78	8602
28.	Babaleshwar	76	8798
29.	Alamatti	76	3021
30.	Sirur	74	8453
31.	Lokapur	72	4138
32.	Galagali	66	4113
33.	Tambe	64	6032
34.	Sulibhavi	64	5881
35.	Savalgi	62	6017
36.	Salotgi	60	7487
37.	Tikota	60	7253
38.	Horti	58	3341
39.	Hunnur	56	5505
40.	Sarwad	54	5252
41.	Ukkali	52	6161
42.	Nidagundi	52	5183
43.	Bevoor	52	4079
44.	Belur	52	3379
45.	Hipparagi	50	6466
46.	Agarkhed	50	5178
47.	Shivanagi	50	4677
48.	Toravi	50	4061

Fifth Order Central Places

49.	There are 52 Fifth order central places ranging between 24 and 48 centrality score values.		

REFERENCES

Berry B.J.L. and Garrison W.M., *The Functional Basis of the Central Place Hierarchy,* Vol.34, 1958, pp.145-154, and pp. 304 to 311.

Brush, J.E., "The Hierarchy of Central Places in South-Western Wisconsin," *Geographical Review*, Vol. 43, 1953, pp. 380-402.

Clarke, P.J., and Evanue, P.C., "Distance to Nearest Neighbour as a Measure of Spatial Relationship in Population," *Ecology*, Vol.35, 1954, pp. 445-453.

Dixit, K.R. amd Diddee, V. "Transactions", *IIG*, Vol. (1) 1979.

Misra, R.P., "Growth Poles and Growth centres in Urban and Regional Planning in India," *Development Studies* No.2, I.D.S., Mysore, 1971.

Pinder, and Withrick, "Principles, Practices and Pitfalls of Nearest Neighbour Analysis, *Geography*, Vol. 57, 1972, pp. 277-288.

19

SPATIAL PLANNING AND INTEGRATED RURAL DEVELOPMENT

SUCHIT SUR

Geography and Spatial Planning

GEOGRAPHY, PARTICULARLY since 1950, has been undergoing certain methodological transformation in the treatment of different elements that have so long been treated from static and also idiographic points of view. The method, though still of some value in academic geography, is unable to make geography or geographers to contribute to the building of the society with purpose.

This methodological transformation with large-scale induction of mathematics or statistics has made possible the development of theories and models, laws and principles that has led to the understanding of the development and distribution of different geographical elements in space, from a scientific point of view. This has focussed attention on processes, particularly on the human behavioural aspects, that concern the evolution and spread of activities over space. Such an emphasis on processes rather than on the description of elements alone has armed geography with a powerful intellectual and technical tool, that is, prediction. No more is the subject solely concentrating its efforts on description but rather on the normative and predictive aspects of any analysis of growth and distribution of elements. This has brought in one salutary effect on the scope of geography, i.e., geographers' involvement in planning — physical, economic and social; national, regional and local.

Space and spatial organisation are concepts that are basic to geography or geographical analysis since the time of Humboldt (1769–1859) and Ritter (1779–1859). Spatial organisation is the sum total and result of spatial process and structure. Structure and process are circularly causal and result in the evolution of particular spatial organisation. "People generate spatial processes in order to satisfy their needs and desires and these processes create spatial structures which in turn influence and modify spatial processes.[1] Activities of people generated by different processes interact with each other depending on the different attributes of places, and 'distance' is the key determining factor in the generation of such processes and evolution and spread of activities in space. Such spatial interaction has three elements — physical distance of time, cost and perception; socio-cultural distance; and functional distance or complementarity. Apart from its economic overtones, the element of complementarity includes much broader aspect of man's social, cultural, political and psychological factors which bring both points and people much closer together and bring down the distance, thereby increasing interactions among them.[2] Conception and analysis of such interrelated elements, and their contributions to the spread and development of human activities in space, demand that geographers should try to rectify spatial incongruities and to take preventive action against possible spatial incompatibility in the future; "understanding and manipulating space and spatial distributions will be the geographers' contribution to human welfare."[3] All these endeavours are termed 'spatial planning' in geography.

Spatial Planning and Integrated Rural Development

In the course of thirty years of planning, Indian plans and policies for narrowing down economic and social disparities among people and space have paid increasing attention to and advocated for progressive diminutive spatial units for planning. Along with this, concepts such as 'integrated area development' at micro-level have become the accepted device and an all-important spatial dimension for comprehensive and integrated development of any micro-level units like community of interest areas, thana, tehsil, taluka, block, sub-division or district at large. Development oriented plans and programmes for rural areas had been taken up since Fourth, and particularly since Fifth Five Year Plans with several target groups and target area approach throughout the country. During the Sixth Five Year Plan (1978-83) and its revision since 1980, greater emphasis has been put on comprehensive and integrated rural development through I.R.D.P. and Block-level planning and ultimately D.R.D.A. In all such plans though major attention is given to the deprived group of small farmers, marginal farmers, share-croppers, landless labourers, artisans and petty traders to pull them up above the poverty line, co-ordination and integration of different programmes and use of local potential resources with a comprehensive rather than individual sectoral development effort are given due importance that were for long found neglected in earlier efforts like the CD Programme or Target Group Programmes. Thus, rural development plans have necessarily become integrated planning at the micro-level with sectoral and spatial integration as the method. Development plans for rural areas are to be prepared in three sequential stages: (a) general service centre plan, (b) sectoral plan, and (c) integrated area development plan. The general service centre plan provides the spatial frame for working on the sectoral proposals for development and finally the functional and spatial integration of such service centres and sectoral activities are given effect to through integrated area development component for a viable community area.[4]

In all these three steps, spatial planning is involved and here geography and geographers can play their role in the development of rural areas. Particularly important is the 'general service centre plan' through which the viable community area for development is chalked out. It also involves 'general settlement and linkage plan' that identifies settlements as focal points at various levels of functional hierarchy for institutional development, and demarcates their hinterlands with analysis of linkages within and outside the area for efficient functioning of the chosen centres, existing or potential, by means of transportation. One important element in planning for rural areas through general settlement and linkage plan is that development and integration are to be based on the notion of 'rural-urban continuum' rather than rural alone as done earlier, e.g. in CD programmes.

In the light of the above discussion an attempt is made for spatial planning of Siliguri sub-division of Darjeeling District of West Bengal.

Spatial Planning for Siliguri Sub-Division

The study area (Fig. 19.1) is one of the four sub-divisions of Darjeeling District and the only sub-division covering the entire plain section of

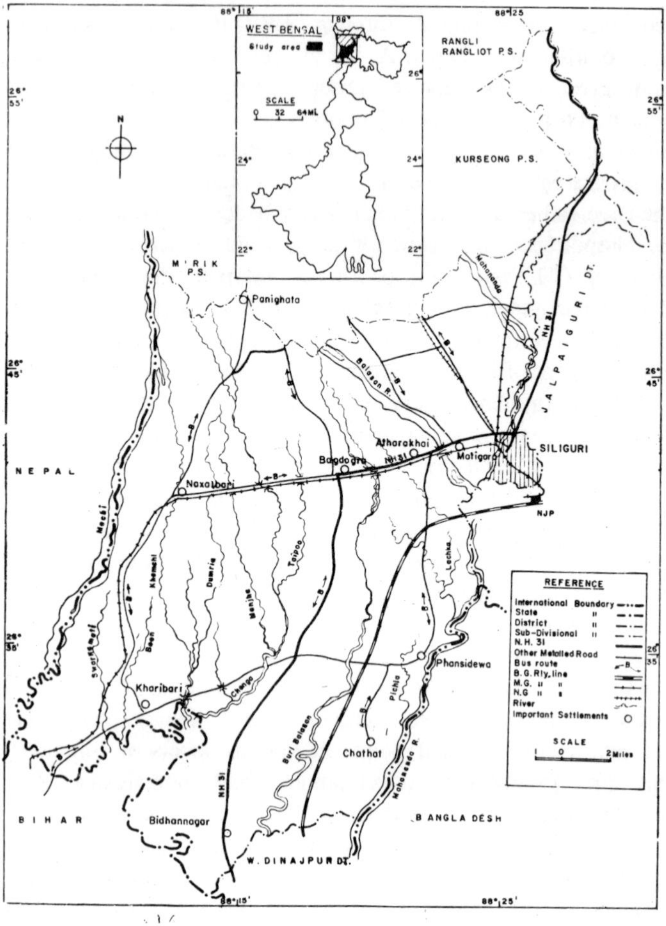

Fig. 19.1: The Study Area: Siliguri-Sub-division

the district. There are two Panchayat Samitis, eighteen Gram Panchayats and four police stations. The total population (excluding Siliguri town) was 2,04,315 in 1971, inhabiting 286 villages and occupying 821.9 sq. km. The density of population on an average was 249 persons per sq. km. Of the total population a large percentage belonged to Scheduled Castes and Scheduled Tribes (40.74%). There are 43 tea gardens covering 46.24 per cent of the total area of the

Sub-division. There is one town and 286 inhabited villages with 83,236 (40.74%) Scheduled Castes and Scheduled Tribes population and 27,326 (38.29%) cultivators.

For identification of focal points in the study area, settlements having periodic hats have been picked up first, as 'periodic *hats*' play most important economic as well as social function for the rural community. But it was found that except for 12 such settlements, other centres possess either no other function or very few, and therefore, were excluded from consideration as focal points. Detailed information were collected both for selected centres as well as for different villages under the jurisdiction of the 18 Gram Panchayats. Altogether 139 functions have been considered covering administrative, extension, trade and commerce, education, health, finance, transport and communication, veterinary and animal husbandry, and services. Centrality score method has been followed for the purpose of evaluation of importance of different focal points.[5] As such, weights of each of 139 functions have been found out with the help of formula $W = 1/f \times 100$ (where f is the total number of functions in the study area and number 100 is the constant for multiplication). Table 19.1 indicates the position of 12 focal points on the basis of centrality scores.

After determination of centrality scores for each settlement the next step was the classification and ranking of settlements. For that purpose, both centrality scores and 'mean and standard deviation methods'[6] were employed and results can be seen from Tables 19.1 and 19.2. It can be seen that of the 12 settlements, one each ranked as class 1, 2 and 3 and the rest nine ranked as class 4. But hierarchy of the settlements, as derived, shows a disproportionate number of both higher and lower order centres in each class which needs rectification through general settlement and linkage plan (Fig. 19.2).

For spatial planning purpose, the next step is to determine different levels of spatial organisation units and respective hinterlands of the focal points. A number of methods are available for the above purpose but none of them can provide a complete solution or can give due consideration to physical, economic and administrative factors which matter most in the spatial interaction by people both in the intra-regional and inter-regional spatial units. Therefore, for the purpose of determination of the First-level of spatial organisation and hinterlands of focal points, following techniques have been employed: (a) desire lines have been drawn for periodic *hats* of each focal point

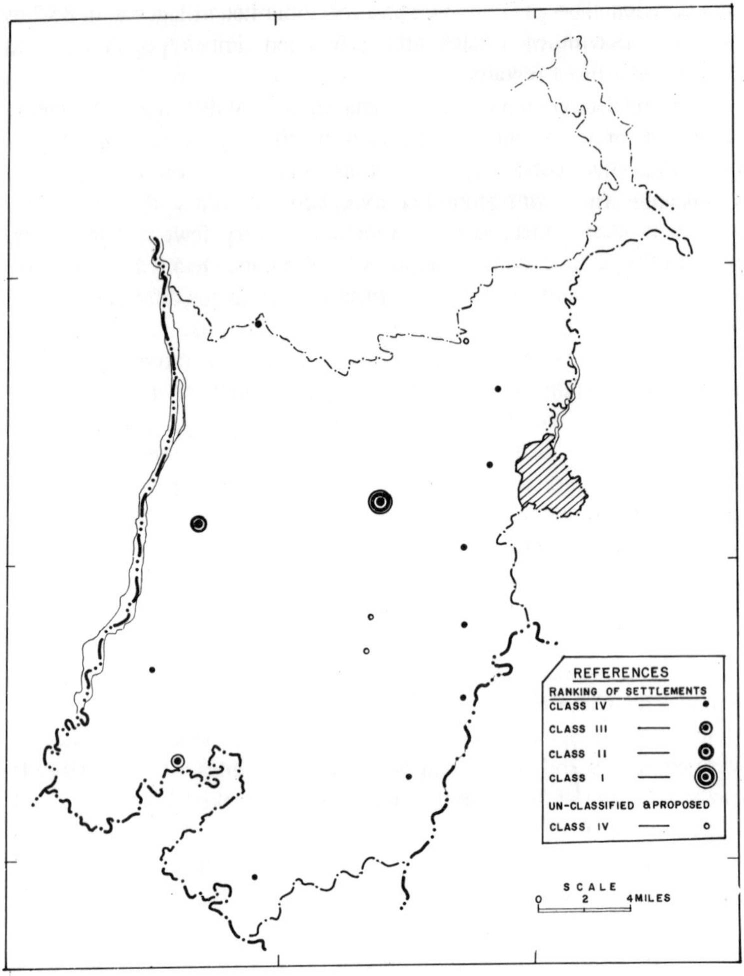

Fig. 19.2: Ranking of Settlements by Mean S.D. Method

and (b) contiguity in space has been adhered to. First-level spatial organisational units and hinterlands of focal points have been determined by chalking out the areas by the application of above two techniques, as at this level spatial interaction is most general and universal. As expected, due to spatio-temporal separation of periodic *hats* and their meetings, there are overlappings in the service area or hinterlands of respective focal points as people of a particular village take advantage of more than one periodic hat in a week or fortnight. While drawing hinterland boundaries, contiguity of villages has been adhered to and distant places without contiguity have been excluded from respective hinterland demarcations. All the 12 focal points as centres of important periodic *hats* for surrounding rural areas, form focal points at the First-level (Fig. 19.3). Table 19.3 presents travel distance norms for different periodic *hats*.

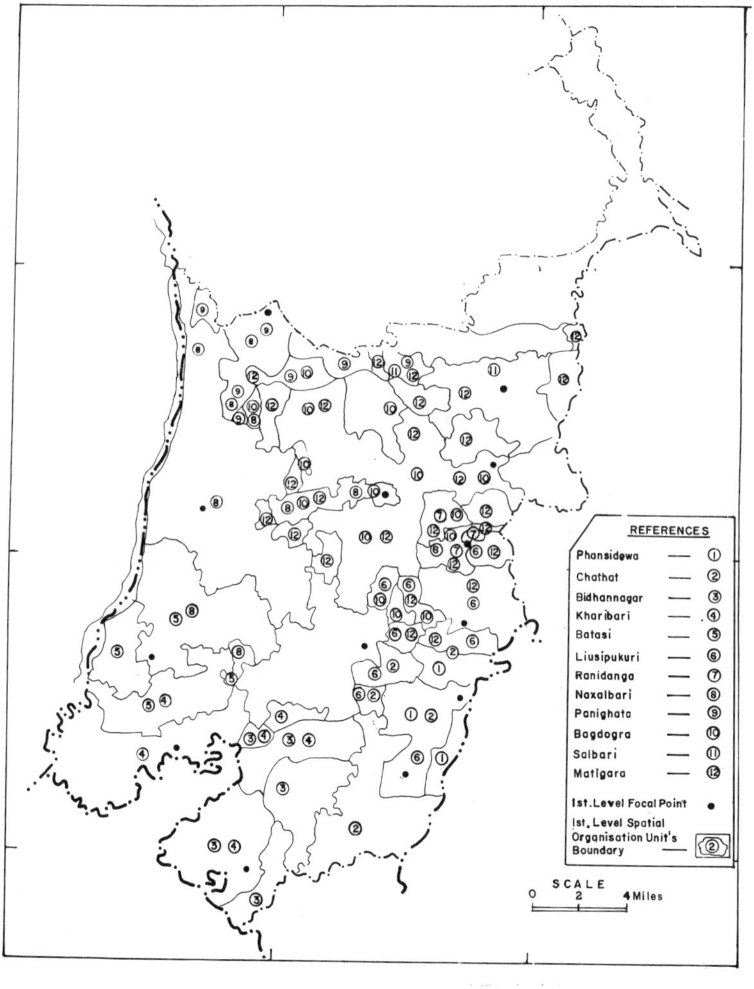

Fig. 19.3: First-Level Spatial Organisational units and Focal Points

TABLE 19.1: Importance of Focal Points

Sl.No.	Name of settlements	Centrality score	Mean & Standard deviation	Class
1.	Bagdogra	2692.10		I
2.	Naxalbari	1578.29		II
3.	Khoribari	743.03		III
			M = 643.57 SD = 753.47	
4.	Phansidewa	578.51		IV
5.	Bidhannagar	487.90		IV
6.	Matigara	455.67		IV
7.	Batasi	276.97		IV
8.	Chathat	261.60		IV
9.	Panighata	236.43		IV
10.	Salbari	211.73		IV
11.	Ranidanga	110.58		IV
12.	Lesupakhri	90.03		IV

Average distance travelled by people is 0.59 miles, which is considered very convenient from the spatial point of view and 75 per cent of them travelled within a distance of 3.99 miles which can be taken as the theoretical distance for measuring the service area around each focal point at the First-level. Table 19.2 shows how those focal points vary in the matter of quality and quantity of services on the basis of which their centrality scores have been calculated.

Determination of Second-level spatial organisational units and demarcation of hinterlands of focal points of that level (Fig. 19.4) have been attempted with the following techniques:

TABLE 19.2: Classification and Ranking of Settlements

Class	Values
I	Between Mean + 3 SD and Mean + 2 SD = Between 643.57 + 3X 733.47 and 643.57 + 2 X 733.47 + Between 2843.98 & 2110.51
II	Between Mean + 2 SD and Mean + 1SD = Between 2110.51 & 1377.04
III	Between Mean + 1 SD and Mean = Between 1377.04 & 643.57
IV	Below Mean = Less than 643.57

(a) more important of the 12 settlements have been singled out for consideration as higher level centres on availability of functions and their quality;

(b) different institutional facilities and their area of operation or jurisdiction have been given due regard, considering people's interaction on that basis. Particular attention has been paid to such institutions as junior land reform office at G.S. level, Gram Panchayat Offices, Commercial and Rural Bank at Local level, and B.D.O. or Panchayat Samiti offices etc. at Block level;

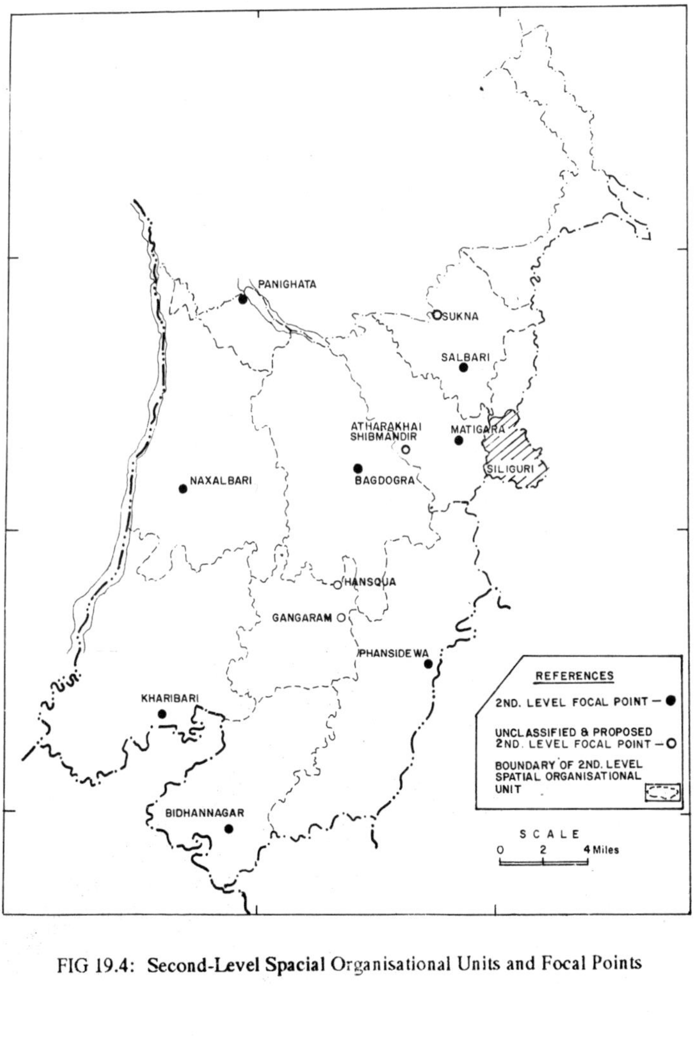

FIG 19.4: **Second-Level Spacial** Organisational Units and Focal Points

TABLE 19.3: **Travel Distance Norms of Periodic Hats of the Study Area**

Distance range in miles	Frequency of persons	Cumulative frequency	% Cumulative frequency	Travel distance norms
0 – 2	1739	1739	43.47	
2 – 4	1262	3001	75.02	
4 – 6	564	3565	89.12	
6 – 8	218	3783	94.57	
8 –10	73	3856	96.40	
10 –12	33	3889	97.22	Average travel distance
12 –14	24	3913	97.82	= 0.59 miles
14 –16	32	3945	98.62	Maximum travel distance
16 –18	22	3967	99.17	= 25.80 miles
18 –20	18	3985	99.67	Minimum travel distance
20 –22	10	3995	99.87	= 0.25 miles
22 –24	0	3995	99.87	75% travelled within a
24 –26	5	4000	100.00	distance of 3.99 miles

(c) physical accessibility and, on that consideration, ease of interaction throughout the year on foot and by conveyances have been taken note of in particular cases; and

(d) contiguity over space has been maintained and villages have been incorporated into separate spatial units following previous three principles as far as practicable.

In order to demarcate these spatial units certain difficulties were faced and to avoid these certain assumptions have been made. As indicated in case of First-level spatial units, all the identified centres do not possess all functions and not all of them are of equal order and quantity (which is usual). But if we had to follow the ranking of settlements, it would have become impossible to demarcate such units and hinterlands of focal points, as only one or two would have emerged as Second-level spatial units and focal points. Such a situation would have betrayed the conditions as mentioned before. Therefore, besides 3 higher level centres — Bagdogra, Naxalbari and Khoribari, which are ranked as class 1, 2 and 3 respectively, other lower order centres have also been considered as belonging to this level of spatial organisation and focal points (Fig. 19.5). The latter settlements — Phansidewa, Bidhannagar, Matigara, Panighata, Salbari and Gangaram — Hansqua — are potential and viable future centres of growth considering their location on the major transportation routes as well as factors like administrative, institutional, banking and physical convenience.

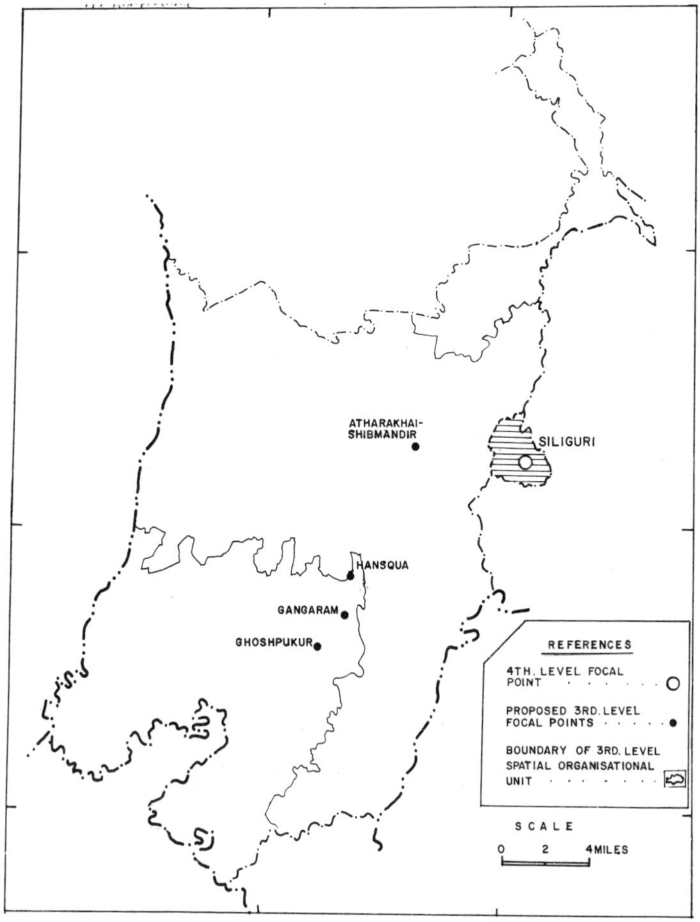

FIG 19.5: Third and Fourth Level Spatial Organisational Units and Focal Points

There are certain other problems in the study area which make clear-cut demarcation impossible. The foremost among the problems is the jurisdiction of Khoribari-Phansidewa Block or *Panchayat Samiti* and location of its headquarters at Phansidewa. The area spreads east-west covering half of the study area. The important fact is that no direct east-west road and bus connection is available. This has resulted in less and below desirable interaction between Phansidewa and its hinterland, and Khoribari located in Phansidewa hinterland is actually ranked higher than Phansidewa itself. Similarly, Bidhannagar, though lying low in the ranking chart, has been considered as a separate and potential growth centre because of its problem of accessibility with Phansidewa in particular and also with Khoribari as it lies on the N.H. 31, connecting Siliguri via Bagdogra by regular bus service. Similar situation prevails in case of villages located in south-central part in the same Phansidewa-Khoribari Block but which do not intensively interact with any of those settlements. This part has one important periodic market centre, Gangaram, but with very few permanent functions. The nearby settlement, Hansqua, may become a centre of growth if government deliberately sets up certain policy functions there as it is in the heart of I.T.D.P area No. 16 of the State. Panighata has been considered as a Second-level centre because of its location away from other centres to the north-western extremity of the study area, serving mainly the contiguous area of the hill sub-division and possessing some functions of interest. Nearness to Siliguri and growth of several important focal points in close juxtaposition has presented another kind of problem for clear-cut demarcation of spatial units and focal points of Bagdogra, Matigara and Salbari. The former two are located on the N.H. 31 connecting Siliguri, and between them is located another growing and important centre — Atharakhai-Shimbmandir — where the Panchayat Samiti Headquarters, University, G.P. Headquarters, High School, etc. are located besides some services and a daily market. Therefore, in the demarcation of spatial units and hiterlands, two centres instead of a single centre with conceivable smaller service areas have been shown in case of Matigara and Salbari, with Sukna possessing one P.H.C. and some services as a second centre with Salbari.

These Second-level spatial units and focal points are to be taken up for planned development through integrated area development at micro-level. Contiguous gram panchayats need to be developed in a coordinated and integrated manner with respect to irrigation, roads and agricultural programme depending on the local variation of relief and resource potentialities with added emphasis on chalking out programmes for development of processing industries and their location at focal points, and promotion of skills and development of industries of artisans. Service and administrative sectors need to work coherently. Much effort on the part of government must be put to develop policy functions in those low ranked centres. Other non-policy functions will automatically come up with their establishment and also with the improvement of the economic condition of the people through rural development programmes already at hand. Government institutions and others, likes S.H. Centres, J.L.R. Office, Veterinary Aid Centres, High Schools etc. will provide various economic and social needs on intra-regional basis, and for their effective utilisation metalled and other roads have to be built and improved upon for settlement linkages.

In the prevailing situation, demarcation of Third-level spatial organisational units and focal points by amalgamation of second-level spatial units and focal points would present difficulties and will involve disregard of local administrative boundaries of different policy functions. But considering the present unwieldy, emotionally incoherent space of Khoribari-Phansidewa C.D. Block, focussing on Phansidewa (as explained in earlier paragraph), certain adjustments in the existing pattern of jurisdiction and boundaries of such policy functions, Panchayat Samiti and C.D. Block Headquarters J.L.R. Office etc., are suggested and proposals are made for the likely development of such newly adjusted boundaries as well as location of new policy functions in potential focal points. Here also, besides contiguity over space with reference to G.P. boundaries, ease of transportation and connectivity and peoples' space preference have been taken into consideration for selecting the Third-level spatial units and focal points. They must be treated as tentative but, from the point of view of development planning, are desirable. Two such spatial units are proposed (Fig. 19.5). The first Third-level spatial organisational unit is formed by the amalgamation of Panighata, Naxalbari, Bagdogra, Matigara, Atharakhai-Shibmandir, Salbari-Sukna and Phansidewa Second-level spatial units and focussed on the existing C.D. Block and *Panchayat Samiti* Headquarters of Atharakhai-Shibmandir (because of administrative, transportation and other complementarity). The second Third-level spatial organisational unit is formed by the amalgamation of Khoribari, Bidhannagar and Gangaram-Hansqua Second-level spatial units and focussed on the proposed new focal point and C.D. Block or *Panchayat Samiti* Headquarters at either Ghoshpukur or Gangaram or Hansqua, for ease of transportation, administration, peoples' space preference etc. This reorganisation is also desirable for stimulating growth impulses in the most backward of the several units of the Second-level — Gangaram-Hansqua and South-western part of Khoribari. Some policy functions have been proposed to be developed at the tentative focal point, e.g., new headquarters of the reorganised C.D. Block, new J.L.R. Office (for Bidhannagar, Ghoshpukur and parts of Hetmuri-Singhijhora G.Ps., new S.H. Centre etc.).

Of the two, the second unit presents agriculture as the single dominant activity with much future potentiality, but contains some pockets of extreme backwardness which require integrated and co-ordinated spatial development planning needing special attention to their employment generation in non-traditional sectors and improvement of the economy of the existing sector. The first unit has both agriculturally prospective and increasingly important non-traditional service segments in the southern and northern parts respectively. The existing unequal development of the two segments is due to more market oriented agricultural activity and increasing opportunities of non-agricultural employment in the northern part for the proximity of Siliguri and nearby urban and semi-urban centres like Bagdogra and Atharakhai-Shibmandir, and potential but traditional agricultural activity in the southern part providing the only employment opportunity.

At this level, the two units are to be developed on the lines of Block-level planning by DRDA. Demarcation of special problem and prospect areas within the units and initiation of special efforts for their development needs are essential for bringing in spatial economic and social integrity. Not only does the existing economic base need to be further strengthened but diversification of opportunities of employment is also imperative in the second unit and the northern part of the first unit. Intra- and inter-regional linkages of settlements is necessary and the existing transport gap between east-west focal points needs immediate bridging (one of the two bridges en-route is under construction and nearing completion) for more articulation of space and rectification of spatial incongruities as well as incompatibility not only at that level but for the entire study area.

Siliguri urban centre, the Sub-Divisional Headquarters and the most important commercial centre and transportational node, forms the fourth and highest level focal point and the entire study area comes within its hinterland.

Conclusion

In conclusion, one important thing must be pointed out that although settlements of different orders can be identified and developed as focal points of growth, their utility in the development spatial planning is less important and the geographical element of 'distance' must count more than what the ranking of settlements may prescribe. Because, unless the socio-cultural distance and several inhibitions on the part of the generally backward population regarding education, health, dairy and poultry, animal husbandry, better and improved practices in agriculture, etc. are narrowed down and neutralised and other social and political organisational rigidities eased out, simple energisation of focal points and their spatial units will not bring the desired filtering of the development processes over the space. They provide the spatial framework and distribution of centres of innovation and spread but backward masses seldom interact with them (except in case of hats) as revealed in this study.

NOTES

1. Abler, Adams & Gould, *Spatial Organisation : The Geographer's View of the World*, Prentice-Hall, New Jersey, 1971.
2. Kostrowicki, Jerzy, A Key Concept : Spatial Organisation, *International Social Science Journal*, Vol. XXVII, NO. 2, 1975.
3. Abler, Adams and Gould, *op. cit.*
4. Saini, N.S., "Planning for Rural Development : An experience in Methodology, *Urban and Rural Planning Thought*,Vol. XIX, No. 3-4, July-Dec., 1976.
5. Thapliyal, B.K., "Spatial Planning for Integrated Area Development in Gazipur and Ballia Districts," *Behavioural and Community Development*, N.I.R.D.
6. Nelson, H.J., "A Service Classification of American Cities," *Economic Geography*, Vol. XXXI, 1965, p 189.

20

CRIMINOGENESIS AND INTEGRATED RURAL DEVELOPMENT
A Case Study of Bhagwanpura

K.K. SHARMA

THERE IS an increasing concern about integrated rural development in India. Large number of empirical studies provide enough evidence to the effect that considerable achievements have been made in that direction.[1,2,3] Several pathologies have arisen as a concomitant and consequence of the process. Criminogenesis is one of them. The purpose of this paper is to test if there is any correlation between the rising criminality and the changing socio-economic conditions in Bhagwanpura village of Ajmer district, and the perceptional response of the villagers.

Theoretical Framework

It is unfortunate that very little attention has been paid to the form of development. It is commonly agreed that the modern-style development follows the Western lines. It appears that the planners have not only been hallucinated by the Westernism but have also been influenced by the Westerners' critical views of our traditional institutions. This shows dominance of the "Western ethnocentrism".[4] However, the purpose of this paper is neither to introduce a new definition of development nor to enter into a controversy of preferring the Western style of development or a form suiting the indigenous structure. Conditions have advanced too far to review. In the process of myopic development aimed at achieving too much too soon, the traditional institutions and values are being dismantled and modern institutions and values, based on an alien mode, are being introduced. This process of Westernisation of the "Sub-cutural Westernisation" is not even wholesome[5] but rather unsystematic and fragmentary. Besides other evils, it has created a gap in social control mechanism, sometimes referred to as a "motivation-organisation vacuum".[6] The main thrusts of the development, economic parity and better standards of living, have remained peripheral. Its benefits have scarcely permeated down to the grass-root,[7,8], while aspirations and material wants are enhanced beyond levels where they can be satisfied.[9] The process, thus, adds dissatisfaction, frustration and intolerance to the simplistic traditional life, with the sting of socio-economic deprivation.[10] Families tend to disintegrated and the command of age as a controlling factor fades, so that the individual is free from much of the scrutiny and control which characterise life in a traditional society. As a result, problems in the area of social peace occur with the development. Propensity to crime is one such problem. Thus a broad thesis shaping this research is that the sources of crime in the process of modern development may not be sought in absolute poverty or wealth, but in relative feelings of discontent, dissatisfaction, family disorganisation, extension of artificial needs and overstimulation of aspirations.

Methodology

A field study of Bhagwanpura village was undertaken. Number of crime incidents, as registered by the police between 1971 and 1980, were collected. Much of the desired socio-economic data were not available for the entire period, thus, limiting the number of explanatory variables to education (number of candidates passing middle standard each year), mean monthly income per family and cases of family disintegration. Perception of all rural family heads was also measured through a questionnaire. It was measured if the criminality, urban effects, family disintegration, education, prosperity, aspirations and frustrations have not increased, slowly increased or fastly increased. These were rated as 0.0, 0.5 and 1.0 respectively.

The first set, thus, provides for a longitudinal experiment, in which the socio-economic variables have been subjected to simple correlation and regression analyses in order to discover the degree and direction of relationships and to obtain predictive equations of the form of $Y=a+bX$.

Analyses

Incidences of crime have increased at an alarming rate during the ten-year period of 1971-80. Figure 20.1 represents the trend clearly. Fairly high correlations of 0.77, 0.7 and 0.82 have been found between criminality and socio-economic variables of education, income and family disintegration respectively. It, however, does not mean that rising levels of education, income and/or family disorganisation directly cause criminogenesis. But these factors may have led to the sociogenesis of such conditions which encouraged criminogenesis.

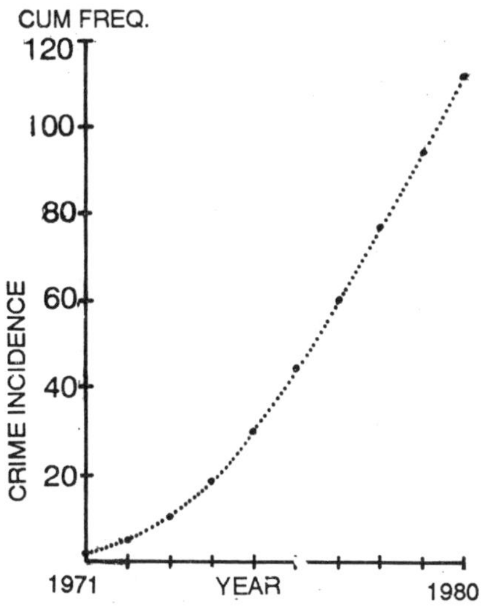

FIG. 20.1: Incidences of Crimes

Relationship between criminogenesis and education had been a part of criminological studies as far back as the 1840s when Fletcher studied the nexus.[11] With a few years of education, students change their self image and their goals to such an extent that they refuse to work on farms because they consider this to be menial labour.[12] As in this case farmers often complain that their sons who receive schooling are of no use to them on the land.[13] They tend to seek office jobs which do not increase in proportion to the "superabundant supply" of educated persons. The result is "the bitter dissatisfaction which education can arouse when it is not matched by economic opportunity".[14] Such education produces socio-structural conditions that create deep feelings of alienation in many youths,[15] so that it is thought of as the "most powerful stimulant to crime".[16] Interpretation of the regression equation, $Y = 2.96 + 1.00X$ (Fig. 20.2), may be valid in the light of such theoretical precepts.

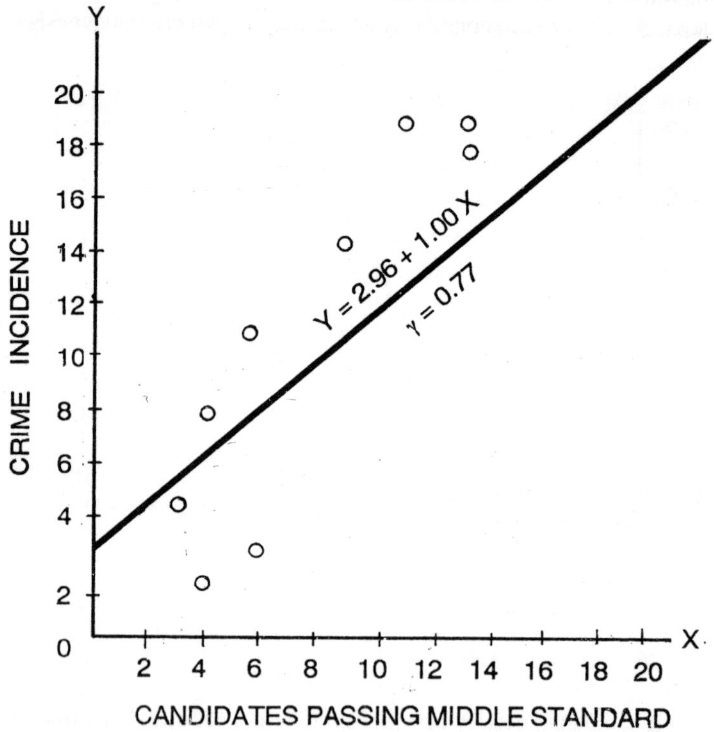

FIG. 20.2: Education Process and Socio-structural Conditions among Youths

As indicated by mean income figures, prosperity has increased with the ongoing development. But the central issue is — "who benefits" most in the process of development. Even though, ideally, development is oriented toward "narrowing the gap between the socio-economically advantaged and disadvantaged segments" of the society,[17] empirical studies dealing with the dynamics of gap generation suggest disproportionately small permeation of benefits to the lowest stratum.[18,19] A recent case study of an Indian village by Sharma (1978) concludes that economic disparities have increased significantly, and this tends to encourage cleavages and class conflicts. In a traditional society, social classes make up a stratification which does not stimulate any attempt for upward mobility. Also never, in either absolute or relative terms, did individuals or families in social classes experience decline or failure in a traditional society. Only with the modern development does that happen due to the transformation of social classes into economic classes in which upward movement is attempted.[20] The observed high correlation between crime and mean income seem to suggest that the consequences of failure in achieving higher status in the world of economic rather than human roles may be pathological. Thus, the regression equation, $Y = -9.5 + 0.05X$ (Fig. 20.3), may be interpreted within this theoretical framework.

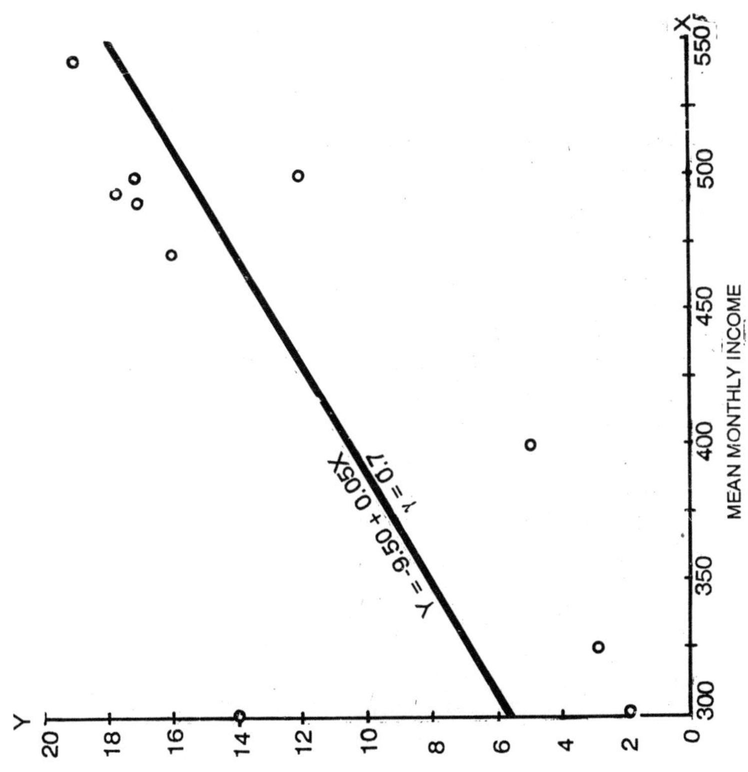

FIG. 20.3: Correlation between Crime and Mean Income

Under the impact of modern development, the extended family begins to disintegrate and is replaced by the nuclear family in which "the tendencies toward distrust and hostility are intensified". The behaviour of one person becomes a matter of relative indifference to others. The family and the neighbourhood are, therefore, no longer competent to control their members, a situation most likely to encourage crime.[21] It is in this perspective that the high correlation between crime and family disintegration becomes significant. The regression equation is obtained in the form of $Y = 3.88 + 2.12X$ (Fig. 20.4).

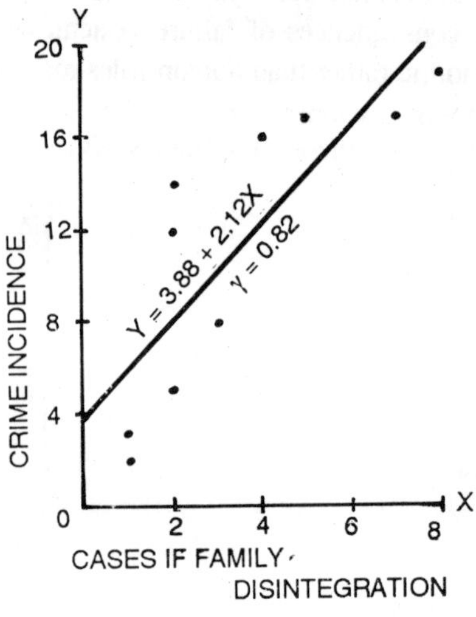

FIG. 20.4: Correlation between Crime and Family Disintegration

Perceptional data forms the second set. Individual perception is highly significant as it provides cognitive measures of mental maps.[22,23] As the villagers perceived, there exists fairly high correlations of criminality with urban effects (0.79), family disintegration (0.63), education (0.66), prosperity (0.52), aspirations (0.7) and frustrations (0.74). All these correlations strengthen theoretical precepts and probable explanations given in the preceding discussion. The perceptional data lend additional knowledge about the respondents' view of the modern development and its effects. The matter is of special significance because not only people's participation but their response is also guided by their cognition of developmental efforts.

Conclusions

Theoretical prepositions of this study have been upheld. The broad thesis that criminality has increased due to certain conditions emerging as concomitant and consequence of the modern development has gained considerable statistical support. However, the thesis needs a rigorous longitudinal and cross-sectional testing.

The empirical results emphasise need for restructuring the educational system to suit the indigenous social structure and the desired goals. Effective measures may be taken to ensure judicious permeation of benefits down to the lowest stratum in order to reduce the gap and the resultant feelings of deprivation and frustration.

Analysis of the perceptional data indicates the level of rural view of the effects of the developmental efforts. The rural mass, therefore, needs to be enlightened about the new system to seek their active participation for successful and speedy change. Otherwise, the consequence of non-participation, reluctance or misperception is disorganisation at all levels — individual, family and societal, which is most likely to encourage criminogenesis. It is, therefore, imperative that social engineers and planners be cautious about such negative aspects of development and provide for preventive measures. Only then, the integrated rural development programmes will remain fully integrated.

NOTES

1. Dutt, A.K. & Noble, A.G., eds, *Rural-Urban Change in India*, Regional Council for International Education, Pittsburgh, 1975.
2. Sharma, M., *The Politics of Inequality : Competition and Control in an Indian Village*, The University Press, Hawai, 1978.
3. Turner, R., ed., *India's Urban Future*, University of California Press, Los Angeles, 1962.
4. Portes, A., "Modernity and Development: A Critique," *Studies in Comparative International Development*, 8 (3), 1973, p 247-79.
5. Singh, Y., *Modernisation of Indian Tradition: A Systematic Study of Social Change*, Thomson Press Ltd., Delhi, 1973.
6. Huntington, S.P., *Political Order in Changing Societies*, Yale University Press, London, 1978.
7. Myrdal, G., *Asian Drama*, The Twentieth Century Fund, New York, 1968.
8. Nayyar, R., "Rural Poverty in Bihar," *The Journal of Development Studies*, 15 (2), 1979, pp 194-201.

9. Roszak, T., "Technocracy's Children," in J.A. Inciardy, eds. *Emerging Social Issues*, Praeger Publishers, New York, 1975.
10. Toby, J., *Affluence and Adolescent Crime, Task Force Report: Juvenile Delinquency and Youth Crime*, The President's Commission on Law Enforcement and Administration of Justice, U.S. Govt. Printing Office, Washington, D.C., 1967.
11. Morris, T., *The Criminal Area: A Study in Social Ecology*, Routledge & Kegan Paul, London, 1957.
12. Dumont, R., *False Start in Africa*, Sphere Books, London, 1966.
13. Zinkin, T., *Challenges in India*, Chatto & Windus, London, 1966.
14. Hunter, G., *Modernising Peasant Societies: A Comparative Study in Asia and Africa*, Oxford University Press, London, 1969, p 12.
15. North, R.C., *The World That Could Be*, Norton & Co., New York, 1976.
16. Ferri, E., *Criminal Sociology*, Translation by J.I. Kelly & J. Lisle, Agathon Press, New York, 1967, p 181.
17. Rogers, E.M., ed., *Communication and Development: Critical Perspectives*, Sage Publications, London, 1976.
18. Haq, M., *The Poverty Curtain: Choices for the Third World*, Columbia University Press, New York, 1976.
19. Islam, N., Social Objectives and the Pakistani Planning, in A. Madison, ed., *Myrdal's Asian Drama: An Interdisciplinary Critique*, Ciriec-Quai de Rome, Liege, Belgium, 1971.
20. Sharma, K.K., *Spatial Aspects of Modernisation and Crime in Rajasthan*, Ph.D. Thesis, Indiana University, Bloomington, 1980.
21. Sheth, H., *Juvenile Delinquency in an Indian Setting*, Popular Book Depot, Bombay, 1961.
22. Gould, P. & Rodney, W., *Mental Maps*, Penguin Books, Middlesex, England, 1974.
23. Lowe, J.C. & Moryadas, S., *The Geography of Movement*, Houghton Mifflin Co., Boston, 1975, Ch. 7.

21

PROBLEMS OF RURAL DEVELOPMENT
A Case of Dangs District in South Gujarat

ANJANA DESAI

'Development' — A Conceptual Problem

THE WORD "development" has been a confusing and a vague term, as it is not understood consistently and clearly by residents, planners and the decision-makers. Development of a country or of a region is often viewed through the lens of economic growth or decline; least importance is given to social and cultural development of residents. Arguments are advanced that, economic factors are measurable and these are taken into consideration, while non-economic variables are not easy to understand and to measure and hence are not taken into consideration. Both the aspects are equally important in the concept of development. An economic growth without social stability and cultural advancement cannot last long; similarly, social stability is dependent on an economic base. As a result, "development" needs to be viewed from both the angles.

Besides the problems of various aspects of development, the issue poses problems of comparisons — both in relative and absolute terms. Since it is not easy to set a norm of development in absolute terms, growth or development is generally measured in relative terms. Problems arise when development is compared over time and space. Demand for goods and commodities changes over time and space at different rates. Things of luxury in one place can be of necessity in another place. It is not very easy to determine the norm of development on a relative scale, as every region or country is at different levels of growth. Moreover, comparisons are not realistic or justified between two situations which are not similar, e.g. rural versus urban areas, industrial sector versus agricultural sector, etc. As a result of comparisons of growth, problems are created when an agricultural area tries to be industrial for the sake of it. The result is that neither the agricultural sector develops nor does the industrial one. The area automatically faces rural poverty and urban congestion through the process of rural-urban migration. It is not meaningful to compare development between rural and urban areas or agricultural and industrial sectors as income generated from rural, agricultural sector is bound to be lower than that of urban industrial output.

Comparisons of development between different environmental situations are also not fair. Environmental and ecological variations are so vast and acute that development cannot be explained or compared between areas easily. Problems of development are bound to differ in arid areas from marshy tracts; hilly forested areas from cyclone hit coastal tracts and so on.

The major hurdle of comparison of development between regions, perhaps, lies in the 'perception' aspect of it. It is rather difficult to ascertain as to how people, planners and decision-makers perceive the growth and development of a region. The problem is that, there is a large gap in the three levels of perception. It often happens that, residents are not aware of development; they accept and get adapted to the poor situation for generations. Planners often see the need of development through their own eyes and come up with plans having not much relevance to the society in question. The decision-makers take decisions without the participation of common mass or planners. It is very difficult to have "development" without the proper linkage of perception of development among residents, planners and decision-makers.

It is quite clear that, the word "development" suffers conceptual problems. The main question is : How to evaluate the process and direction of development over time and space? Perhaps the root problem lies in copying 'development' of one area for another one. Such an operation not only makes plans irrelevant to social needs but also puts a country always in a dependent situation. Developmental programmes need to be outlined in terms of priority of basic needs of food, shelter, clothes and education so that, the disparity between regions can be reduced and then other progressive plans can be made based on the needs of the society.

Problems of Measuring "Development"

The term 'development' does not only suffer from conceptual problems but also from measurement problems. The two main aspects of development can be economic and socio-cultural. Economic variables that are normally considered, are gross national product, total and per capita income, per capita consumption of power and commodities, rate of economic growth, purchasing power and saving capabilities of residents. The availability of such detailed economic data on an individual levels is rather low. Compatibility and consistency of economic data over space and time are other aspects of the problems.

Since development of any area is not just an economic one — it should include the socio-cultural aspects also. The main hurdle of socio-cultural variables lies in measurements as well as availability of data. The problem is a genuine one when one wants to measure the 'perception' of development. The need and the type of development are not the same for all regions, as the perception of growth varies over time, space and the people concerned. The government authority may like to develop an area as a political power; another government may put resources for economic or social development. One area may be planned to develop only viable growth centres; another area may be designed to develop evenly by removing the socio-economic disparity over space. Perception of government differs — whether to develop rural areas or urban centres, whether to develop industries or agriculture. Since perceptions vary, there may not be any consistency in planning 'development', if there is any change at the government level. Similarly, there is no consistency or perception of development among rich and poor sections of the society. As a result, measurement of perception remains to be a problem.

In short, due to the problem of measuring non-economic variables, they are often omitted from assessment in the process of planning. As a result, development plans are often not relevant to the society.

Problems of Implementing "Development"

The question of implementation poses another aspect of problems of rural development. Besides the changing policy of the government, the question remains vague : who will develop rural areas? Is it the government, private enterprises, planners, social workers or residents?

Private organisations do not like to take risk in setting up industries or business in rural areas because of the low purchasing power of residents and low demand for goods and commodities. It is easier for the government to take up such risks. The government can take care of economic aspects. What about the social aspects and life style of residents? Perhaps the social workers can reach the residents in a better way by bringing awareness among residents in terms of both social and economic development of the area. Awareness lies in residents' perception of development. The question is : are residents concerned? Will they participate in development programmes? It is obvious, any developmental plan requires the full participation of the government, private enterprises, social workers and residents. Due to lack of co-ordination and sincere efforts, implementation continues to be a problem, however good the plans may be. Moreover, limited resources, small funds and poor management are often responsible for low-level of implementation of development programmes, especially in rural areas.

The Dangs — Its Characteristics

The district of Dangs stands out not only as a political unit but also as a region with special physical and human characteristics with its tribal population in a totally rural, hilly and forested environment. It is the second smallest district in Gujarat, occupying an area of 1683 sq. km and is located in the south-eastern part of the state.

The terrain is hilly and undulating, the height varying between 105 and 1317 metres which gradually increases from west to east, near the margin of Maharashtra. Since the annual rainfall is on the heavy side, varying between 100–200 cm, the hills are covered with fairly dense forest of teak, bamboo and other deciduous trees. Soil erosion is prevalent due to heavy rains and surface run-off.

The district is an economically depressed area with very low income of residents, low production of food crop and inadequate amenities and facilities for growth. Medical and educational facilities are in a very poor condition (Table 21.1). Only four per cent of the villages have electricity. The majority of the population depends on wells or rivers for drinking water. The problem of water is acute, especially in summer, when rivers and wells get dried up. The overall life-style of the tribals depends on natural environment with bare minimum for food, clothing and shelter.

TABLE 21.1: Amenities and Facilities in Dangs (1971)

	Amenities	No. of villages having amenities	Percentage
1.	Primary Schools	278	89.00
2.	Secondary Schools	2	0.64
3.	Dispensary with hospital	3	0.96
4.	Electricity	13	4.18
5.	Tap water supply	2	0.64
	Total villages	311	100.00

Lack of interaction is the predominant feature within and outside the district. Residents are "spatially restricted" and their life is limited to a local area. Distance is a barrier here and each village tends to be independent.

Ahwa is the District Headquarters, while Waghai, on the western side, is the main commercial centre. Pimpri and Saputara are other important villages with electricity, school, weekly market, post office and connection with metalled roads. Both transportation and communication networks are rather poor in the district (Table 21.2). A small portion of the district is connected with a narrow-gauge railway up to Waghai.

TABLE 21.2: Connectivity Network in Dangs (1971)

	Connectivity	No. of villages	Percentage
1.	Bus service (all seasons)	43	13.82
2.	Bus services (fair season)	69	22.18
3.	Metalled road	84	27.00
4.	Post office	35	11.25
5.	Posts and Telegraphs	2	0.64
6.	Telephone	3	0.96
7.	Narrow gauge railway (up to Waghai)	1	0.32
	Total villages	311	100.00

Source: *Census of India, District Hand Book*, 1971.

Demographically the Dangs is a unique district with 94 per cent of population as tribals — Bhils and Konkanis. The district is thinly populated. The entire area is rural with residents having a low-level of education, compared to that of Gujarat as a whole (Table 21.3).

Of the two tribes, the Bhils are economically poorer. This is reflected in their fewer material possessions, poorer type of house and its maintenance, lower yield of crop per acre and annual income per family and the type of labour used on the land.*

The impact of environment is clearly noticeable on the economic life of the people — how land is utilised, type of fertiliser used, techniques used in farming, water supply and the type of crops produced in relation to soil, height of the land and the availability of water. Nearly 65 per cent of the residents are farmers and 20 per cent are agricultural labourers.

TABLE 21.3: **Demographic Characteristics of Dangs in Relation to Gujarat (1971)**

Sl. No.	Characteristics	Gujarat	Dangs
1.	Total Population (1971)	26,697,475	94,185
	Males	13,802,494	48,394
	Females	12,894,981	45,791
2.	Tribal Population	13%	94%
3.	Rural Population	19,200,975	94,185
4.	Urban Population	7,496,011	Nil
5.	Density per sq. km	136	56
6.	Sex ratio	934	946
7.	Literacy rate :	35.79	14.16
	Males	46.11	20.80
	Females	24.75	7.13
8.	Growth rate of Population	+29.39	+31.60

Source: *Census of India, District Hand Book*, 1971.

Though the district has enough rainfall, land is cultivated once a year with a low yield per acre. There is an acute shortage of water during summer. High rate of surface run-off has made the top soil infertile due to erosion. Lack of levelled land is also responsible for low yield per acre. The techniques of farming are age-old : cow-dung and burnt leaves are used for fertiliser, rain water is the main source of water for cultivation. A combination of shifting and sedentary cultivation is practised. A portion of the land is often kept fallow, or multiple cropping is practised to retain land-fertility. The agricultural plots are rather small — nearly 46 per cent of the plots are less than one acre and hence total production per family is very low. It is

* Field work was carried out in a village, Hanuvant Chond, near Pimpri, where both Bhils and Konkanis live.

annually 50 kg per family for the Bhils and 400 kg for the Konkanis with larger plots. The main crop is Nagli — a type of low quality millet.

Since the return from the land is poor, the tribals depend on other jobs as temporary labour in cutting bamboo and sugarcane or loading and unloading of logs on trucks. The tribals are also engaged in food gathering, pasturing, fishing and hunting.

Poverty is acute in the Dangs. Most tribals do not get two meals a day and the food is of very low nutritional value. Their income is very low and varies from month to month. The average annual income of Bhils is Rs. 200 per family and Rs. 900 for Konkani families. Income is so low that, the tribals carry on exchange on barter system with the help of grains.

In short, the life-style in Dangs is heavily dependent on natural environment and is below subsistence level. Food with low nutritional value and drinking habits have made them subdued in nature and they seem to have accepted poverty as their life-style. Man here appears to be a part of nature rather than a transformer of it.

Barriers to Growth in the Dangs

The district of Dangs poses a unique problem of rural development. The region has the wealth of forest, yet the people are very poor. The area enjoys heavy rainfall, but it suffers from low yield and acute shortage of water in summer. The region is close to urban and industrial centres like Surat and Nasik, yet it has remained without much economic interaction. There is acute poverty, yet the people have shown no urge to migrate to other places for seeking job opportunities. People are happy in natural surroundings in spite of poverty. It is important that, development programmes are perceived through the eyes of the tribals so that they become relevant to their lives and society. A comprehensive development programme is necessary, covering both the economic and social aspects of life. Socio-economic problems of Dangs are physical as well as human. A number of factors are responsible for it : environmental, spatial, economic, social and cultural, which are discussed below.

Agricultural problems are closely related to environmental situations : lack of level land, lack of water supply in summer and of good fertile top soil.

Inaccessibility, poor exposure and lack of competition have made wages at Dangs rather low — lower than the minimum rate of wages.

The district does not have any significant growth centres. As a result, the region does not appear to be a potential area for growth.

Economic factors are responsible for the slow growth of Dangs. Yield per acre is poor due to poor techniques used in agriculture, low level of farming skill, small plots for cultivation, limited use of good fertiliser and seeds and very poor managements of soil and water resources. Forest wealth is not utilised in the district. It is exported to other parts of the country without providing job-opportunities to the local people. There is no subsidiary job for residents along with part-time farming. As a result the purchasing power of residents is very low.

Social factors are equally responsible for the lack of development among Dangs. Low level of education is associated with poor skill and low income of the tribals. They are not socially organised to bargain for minimum wages. Life is based on day-to-day living, they have no idea of savings and investment for the future. If there is extra money, it is used up for drinking and merry-making.

Culturally, the tribals have accepted poverty in their lives over a period of time. Children are born and brought up in poverty and their needs for a better life is limited. Unless the perception of life is changed for a better one, tribals will not be able to get out of the vicious circle of poverty.

Conclusions and Reflections

A number of measures can be suggested to improve the living conditions of the Dangs : (a) environmental programmes in soil, slope and water management; (b) spatial management by removing seclusion and inaccessibility; (c) creation of job opportunities in small scale, forest based industries, poultry and sheep rearing, and (d) increasing awareness and education among the tribals to have better standard of living.

Development may not be copied or imposed. Every area is a unique one and has its own problems. Problems of the Dangs need to be identified in its spatial context. For their solution comprehensive development plans — covering both social and economic aspects of life, should be chalked out with public participation. It is necessary that, the development programmes be formulated in close collaboration with the tribals. They should figure out the priorities in relation to funds and resources available to them and then it can be implemented through voluntary efforts of the tribals in the light of their social needs.

INDEX